13760598

PRESERVATION OF HISTORICAL RECORDS

Declaration of Independence, U.S. Constitution, and Bill of Rights on display at the National Archives. Meticulous care is taken to preserve these documents for future generations.

PRESERVATION OF HISTORICAL RECORDS

Committee on Preservation of Historical Records
National Materials Advisory Board
Commission on Engineering and Technical Systems
National Research Council

National Academy Press
Washington, D.C.   1986

NATIONAL ACADEMY PRESS   2101 CONSTITUTION AVE., NW   WASHINGTON, DC 20418

This study was conducted under contract GS-00N-84-DSC-0010 between the General Services Administration, National Archives and Records Service, and the National Academy of Sciences. This is National Materials Advisory Board publication NMAB-432.

Library of Congress Cataloging in Publication Data

Preservation of historical records.

Includes index.
1. Manuscripts—Conservation and restoration.
2. Archival materials—Conservation and restoration.
3. History—Sources—Conservation and restoration.
I. National Research Council (U.S.). Committee on Preservation of Historical Records.
Z110.C7P84   1986   025.7   86-12718

ISBN 0-309-03681-X

Committee on
Preservation of Historical Records

Chairman

PETER Z. ADELSTEIN, Materials Sciences and Engineering Division, Eastman Kodak Company, Rochester, New York

Members

GLEN R. CASS, Department of Environmental Engineering, California Institute of Technology, Pasadena, California

HANS H. G. JELLINEK, Chemistry Department, Clarkson University, Potsdam, New York

LEON KATZ, James River Corporation, Norwalk, Connecticut

GEORGE B. KELLY, JR., Consultant in Paper Chemicals, Gaithersburg, Maryland

JOHN C. MALLINSON, Center for Magnetic Recording Research, University of California, San Diego, California

ERNEST R. MAY, Kennedy School of Government, Harvard University, Cambridge, Massachusetts

TERRY O. NORRIS, Nekoosa Papers, Inc., Port Edwards, Wisconsin

TED F. POWELL, Micrographics Division, Genealogical Society of Utah, Salt Lake City, Utah

KWAN Y. WONG, IBM Research Laboratory, San Jose, California

FRANCIS T. S. YU, Department of Electrical Engineering, Pennsylvania State University, University Park, Pennsylvania

Technical Advisor

NORBERT S. BAER, Institute of Fine Arts, New York University, New York, New York

Liaison Representatives

ALAN R. CALMES, National Archives and Records Administration, Washington, D.C.

JOHN C. DAVIS, National Security Agency, Fort Meade, Maryland

KEITH R. EBERHARDT, National Bureau of Standards, Gaithersburg, Maryland

DAVID J. E. SAUMWEBER, NAS Archives, National Academy of Sciences, Washington, D.C.

LESLIE E. SMITH, National Bureau of Standards, Gaithersburg, Maryland

PETER G. WATERS, Library of Congress, Washington, D.C.

NMAB Staff

GEORGE ECONOMOS, *Staff Officer*

C. L. STEELE, *Senior Secretary*

National Materials Advisory Board

Foreword

Contemporary scholars dealing with the past face a paradoxical reversal of conditions. Until very recently, historians worried about the scarcity of sources. Each bit of writing, a unique record from the past, had intrinsic value; and much of the historian's task required the tracking down of a unique manuscript and the recovery of incomplete files. But to compensate, the surviving documents were durable—whether inscribed on clay tablets or written on parchment, vellum, or rag paper. Resisting deterioration, they came down through the centuries intact, many being almost as legible as when written or printed. The custodians of these materials had a comparatively simple task. They could preserve with relatively little difficulty what the accidents of time had randomly selected.

Twentieth-century conditions reverse those of the past. The volume of materials is immense—3 billion items in the National Archives alone, and as many more in state and local archives, in historical societies, and in libraries. The selection of what to retain and what to discard is a complex process. But, by contrast with the past, the materials themselves are fragile; whether comprised of paper made from pulp in the past century, or tape, or disks, they are subject to eventual deterioration due to such factors as humidity and heat as well as frequency of handling. In the absence of energetic preservation programs, these valuable resources for understanding the past will crumble away.

The task is formidable, and the distinguished committee of experts that addressed it in the book which follows has produced an enlightening analysis of the problem and sensible recommendations for its solution. The most important of these are the use of improved paper in the original government records and the establishment of standards for preservation of nonprinted materials on tape and disks.

The scholarly community will await the implementation of these recommendations and the collaboration of the National Archives with other agencies in establishing procedures for the preservation of the treasures they hold.

OSCAR HANDLIN
Carl M. Loeb University Professor
Harvard University

Preface

The National Archives and Records Administration (NARA) is the final repository for permanently valuable federal government documents. The retention, preservation, and possible disposal of documents are the responsibility of NARA. Today, the immensity of this assigned task, which involves some 3 billion pieces of paper, has led NARA to seek independent guidance because of the economic, technical, social, and political implications of its actions. The National Academy of Sciences was called on in 1880 and 1903 to make recommendations regarding the preservation of the Declaration of Independence, and again in 1975 to study the preservation of documents in the event of a nuclear attack. The request for this present study is a recognition of the fact that technically well-founded and realistic advice is needed on the preservation of information contained in paper-based records whose originals may not be intrinsically valuable.

To arrest the deterioration of existing paper documents and to preserve those in an advanced state of degradation, processing treatments and image transfer techniques offer promise if based on adequate economic and technical background data. Much of the latter was available within the committee, and additional pertinent information was obtained from invited guest contributors and other technical experts.

PETER Z. ADELSTEIN
Chairman

Acknowledgments

The Committee on Preservation of Historical Records is grateful to a large number of individuals for their contributions to its data collection and assessment efforts:

H. NEAL BERTRAM, University of California, San Diego—Longevity of magnetic tapes

MILBURN M. COCHRAN, IBM Corporation, Tuscon—Field experience with magnetic tape recordings

GLEN R. CASS, California Institute of Technology—Environmental pollutant criteria for storage

EDWARD F. CUDDIHY, Jet Propulsion Laboratory—Magnetic tape as a storage medium

FRANKLYN E. DAILEY, Image Technology and Applications—Automated image transfer techniques

DAVID H. DAVIES, 3M Company—Long-term stability of optical disks

JOHN C. DAVIS, National Security Agency—Optical disk data storage

KEITH R. EBERHARDT, National Bureau of Standards—Effects of an archives indexing system

EDMUND L. GRAMINSKI, U.S. Bureau of Engraving and Printing—Effectiveness of deacidification

HANS H. G. JELLINEK, Clarkson University—Cellulose chemistry

JOHN B. KELLY, consultant on paper chemicals—Deacidification processes

VINCENT D. LEE-THORP, Lee-Thorp, Inc.—Design options for environmental controls for the Archives building

JOHN C. MALLINSON, University of California, San Diego—Developments in magnetic tape research

ERNEST R. MAY, Harvard University—Experiences in library research

TED F. POWELL, Genealogical Society of Utah—Experiences and practices of microfilming

JOSEPH W. PRICE, Library of Congress—Optical disks as a storage medium

DAVID J. E. SAUMWEBER, NAS Archives—A matrix chart for paper preservation actions

RALPH E. SCHOFER, U.S. Bureau of Standards—Cost-benefit analysis of transferring paper records

LESLIE E. SMITH, National Bureau of Standards—Deterioration of magnetic tape

JOHN F. WATERHOUSE, Institute of Paper Chemistry—Research on long-term-stability paper

KWAN Y. WONG, IBM Corporation, San Jose—(1) Costs of copying; (2) the applicability of semiconductors to archival storage

JAMES E. WOODS, Honeywell Physical Sciences Center—Capabilities of heating, ventilating, and air conditioning systems

FRANCIS T. S. YU, Pennsylvania State University—An optical color signal processing technique

The committee thanks the government liaison representatives for supplying valuable background information and guidance and for assisting the committee in compiling up-to-date information on areas such as optical disks, polymer and paper chemistry, statistical analysis, and archival problems and actions.

Special thanks go to Norbert S. Baer, chairman of NARA's Advisory Committee on Preservation, who supplied technical guidance to the committee throughout the study. Anna K. Nelson, Project Director for the American Council of Learned Societies Committee on the Records of Government (chaired by committee member Ernest R. May), generously provided copies of that committee's report, which were used in this group's deliberation.

Committee chairman Peter Z. Adelstein wishes to give special acknowledgment to committee members who coordinated information for various chapters and prepared the final text. Finally, the chairman thanks the entire committee for its patience during the extensive reviews of the various issues addressed and for assembling the pertinent facts in an open-minded and professional manner.

Abstract

The National Archives is the repository for permanently valuable documents of various federal government agencies. Concerns exist about the condition of some stored paper records because of the increasing quantities that must be handled and the deteriorating condition of some of them. Various methods for preserving paper records were examined, and alternative actions for preserving the original documents or retaining more permanently the information contained in them were assessed. The accessibility requirements of the Archives retrieval system limit the acceptable preservation alternatives for most of the at-risk holdings to photocopying and photographic film storage. Environmental effects are discussed, and standards for potentially dangerous airborne contaminants in the Archives storage areas are developed. Continued monitoring of potentially high-risk records is necessary so that timely corrective action can be taken.

Contents

WHAT IS PAST
IS PROLOGUE

*Sculpture by Robert Aitken at entrance to the National Archives.
The epigraph is from Act II of Shakespeare's* The Tempest.

Preservation of Historical Records

Constitution Avenue entrance to the National Archives Building. In addition to famous documents, millions of other historical records are stored there.

1

Recommendations

The recommendations of the Committee on Preservation of Historical Records on various options available to the National Archives are presented below. Some general recommendations are given first, and these are followed in turn by recommendations on mass treatment of records, on archival copying, on preliminary preservation actions, and on preservation strategy.

GENERAL RECOMMENDATIONS

1. A general improvement in the quality of paper used by the federal government would be an important step in minimizing future paper problems of the sort now experienced by the National Archives. Since permanent papers are available at a reasonable cost, the government should use such papers for records that have permanent value.

2. Archival standards are available for papers and photographic films. Archival standards are also available for electrophotographic reproduction. The National Archives and Records Administration (NARA) should ensure that the records it creates or copies with these media or processes meet these standards.

3. Archival standards do not exist for magnetic tape or optical disks or for the reproduction of records on such media. Since these media are currently being used by the federal government, and since their use will greatly expand in the future, NARA should promote the development of standards for these media at the earliest possible date.

4. Major deposits of machine-readable records exist. If these records are to be useful to future research at the National Archives, NARA should be prepared to accession them and to preserve the information they contain.

RECOMMENDATIONS ON MASS TREATMENT

5. The standards given in Chapter 3 for temperature, humidity, and pollutants should be implemented (see Tables 3-4 and 3-5 for specific standards).

6. NARA should conduct a study of archival storage containers and microenvironments, including boxes, folders, and polyester encapsulation, with a view to understanding the maximum benefit that can be obtained from particular materials and designs. The committee feels that this is an underexplored area that may yield results highly significant to NARA's preservation efforts.

7. NARA should not undertake a mass deacidification program at this time but should monitor the development of deacidification processes.

RECOMMENDATIONS ON ARCHIVAL COPYING

8. The media that are appropriate for archival preservation are paper and photographic film, and the processes appropriate to copying using these media are archivally standard electrophotographic processes (for paper) and silver-based micrographic processes (for film).

9. The materials and technical problems inherent in the use of magnetic and optical storage media and the lack of suitable standards for archival quality make their use as preservation media for archival storage inappropriate at the present time.

RECOMMENDATIONS FOR PRESERVATION ACTION

10. The National Archives should institute procedures immediately that will yield statistics concerning damaged records that are useful for long-term preservation planning and for deciding treatment priorities. The committee suggests that statistics be kept that reflect the condition of records used both in the reading room and by the staff, and that these statistics be supplemented by the information generated by the ongoing maintenance operations.

11. The National Archives should establish criteria for frequent and infrequent use and for satisfactory and unsatisfactory conditions so that priorities in treatment may be assigned.

RECOMMENDATIONS ON PRESERVATION STRATEGY

12. The National Archives should adopt the decision procedure and the recommendations on treatment and records disposal that are embodied in the decision tree, Figure 8-1, with the caution that this recommendation cannot be separated from Recommendation 13.

13. Portions of the proposed preservation plan include disposal of original records after copying. In these cases, the copy *will be* the record. The National Archives must establish in perpetuity a program of effective quality control and verification of copying.

National Archives Central Research Room. A wide range of historical records is available for public examination.

Record storage stacks at the National Archives. The huge collection includes bound volumes, boxed papers, and paper records of all shapes and sizes.

2

Introduction

CURRENT SITUATION AT THE NATIONAL ARCHIVES

The National Archives and Records Service (NARS)* was organized in 1934. Permanently valuable documents that were in storage in various agencies of the federal government were collected to form the holdings of the National Archives. These holdings have been increasing rapidly ever since. The quantity of documents is vast and consists mostly of paper records of various shapes, sizes, and physical conditions. Today there are over 3 billion pieces of paper (Calmes et al., 1985), and this figure is increasing at the rate of close to 3 percent annually. Tremendous archiving problems have been created because of both the volume of material to be stored and the loss of stability of large quantities of paper-based records already in storage. State governments have similar archiving problems (National Association of Government Archives and Records Administrators, 1986), which, although not specifically discussed in this report, appear to be comparably urgent.

Many documents are over 150 years old, and a large number have deteriorated significantly. About a half-billion pages of historical information are at a very high risk of being lost (see Table 2-1). A significant percentage of the total collection consists of modern papers rather than older records. For example, rapid action is required to prevent information loss from "quick-copy" reproductions such as stencil, Mimeograph, and Thermofax produced during the 1940s, 1950s, and 1960s. It is estimated that there are 270 million sheets of paper in this category alone. A pilot survey of National Archives documents estimated that 98.8 percent are quite acidic, with a pH less than 4.5 (Calmes et al., 1982). This is of concern because of the known higher degradation rate of acidic paper.

---

*In April 1985 the National Archives and Records Service (NARS) of the General Services Administration became an independent agency, renamed National Archives and Records Administration (NARA). In this report, the agency is referred to as the National Archives or NARA.

5

TABLE 2-1   Survey of High-Risk Paper Documents

Apparent Condition	Approximate Number of Pages
Already suffered major damage	160 million
Subject to damage by frequent handling	100 million
Deteriorating "quick-copy" reproductions of the 1940s, 1950s, and 1960s	270 million
Total at high risk of loss	530 million

SOURCE: Calmes et al., 1985.

In 2 decades, the Archives holdings will grow to well over 4 billion sheets. Since NARA usually receives papers many years after they have been prepared or collected by the agencies, many of these papers may already have deteriorated. This will increase the numbers given in Table 2-1.

Materials must be preserved, and those in an advanced state of deterioration must be treated to retard further degradation or must be copied. If they are to be copied, a decision must be made as to whether they should be copied onto paper, film, magnetic tape, or optical disk. In addition, a retention policy must be formulated for handling original materials that are to be copied for which the originals do not have any intrinsic value* (National Archives and Records Service, 1982). It is estimated that about 75 percent of the documents have no intrinsic value—that is, they need not be retained in their original form to preserve the information (Calmes et al., 1985). The problem is large because of the large number of paper documents involved.

NARA should take action, and the key question is what action should be taken. Policies developed must be based on the best technical input that is available today. To obtain such guidance, NARA requested an independent study by the National Materials Advisory Board of the National Research Council, and a committee was appointed to examine the options for action. The committee was composed of individuals having the wide background and experience required for this study (see Appendix C).

SCOPE OF STUDY

The specific scope of the committee's inquiry, as defined by NARA, was to make recommendations on how to handle original paper records and on the advisability of transferring information from original paper records to media having acceptable permanence, including media with limited life but capable of being recopied. The committee was also to make recommendations on the disposition of the original document if the information is copied. In the course of its deliberations, the committee considered some important issues that will affect the preservation of future holdings:

1. Only paper records in current holdings were to be considered. Nontextual items, such as motion pictures, photographs, sound recordings, and machine-

---

*Specialized terms are defined in Appendix B.

*Document conservation laboratory at the National Archives. Deteriorated
paper records are treated to slow the ravages of time and handling.*

readable records, were outside the scope of this report, as were such highly specialized materials as maps, artistic prints and drawings, large ledgers, and small diaries.

2. Only paper having no intrinsic value was to be included in the recommendations for transfer of information from the original documents and disposition of the latter.

3. NARA's prime function is to preserve the documentary heritage of the United States and to provide timely service to both government and scholarly researchers (Calmes et al., 1985). In most cases rapid access is not of prime importance at the National Archives and was not considered by the committee. Preservation, therefore, is the primary goal.

OUTLINE OF PROPOSED ACTIONS

This report is divided into chapters that discuss the relative advantages and disadvantages of particular approaches. Several of these proposed actions are not mutually exclusive and can be done concurrently.

Environmental Considerations

Chapter 3 covers the effects of temperature, relative humidity, and pollutants, particularly as they relate to the National Archives. Pollutant levels, removal systems, and recommended standards are covered. Both the benefits to be gained and the problems associated with environmental changes must be considered. A key question is to what degree environmental changes alone, or in combination with other approaches, reduce the threat of deterioration. It is believed that many documents, even on moderately acid paper, that are very infrequently used deteriorate slowly if they are properly housed and stored.

Paper

Chapters 4 through 7 discuss the properties of various materials that can be used for copying deteriorated documents. Paper is of greatest interest since it not only is a copying material but also comprises the largest holdings by far of the National Archives. Thus, Chapter 4 gives a detailed discussion of the manufacture and behavior of paper. The discussion also covers copying deteriorated documents onto more permanent paper or treating the existing documents to prolong their life, or both. Considerations for copying are the life expectancy of currently available papers, the permanency of inks and toners, and the advantages and disadvantages of storing paper documents. Treatment of existing documents involves the benefits and problems associated with the various deacidification processes.

Photographic Film

Chapter 5 covers the types of photographic film available and their life expectancies. Advantages are the predictable long life, high storage capacity, human-readable form, and the existence of recognized standards. Disadvantages are film's susceptibility to damage from adverse storage conditions, cost of verification of the copying process, and low direct manipulability.

Magnetic Recording Media

Considerations for magnetic recording media discussed in Chapter 6 include the stability of magnetic materials, the chemical stability of the component layers, and the mechanical stability of the tape roll. Problems associated with wear and the continued availability of appropriate hardware and software are critical factors. Advantages include high data storage density and data manipulability.

Optical Disks

The archival quality of optical disks is addressed in Chapter 7, as is the practicality of continued recopying should disks degrade. The advantages of data compaction and access time must be balanced against the problems of the archival life of the hardware and software.

Semiconductors

The feasibility of storing information on semiconductors is briefly addressed in Appendix A.

Discussion of Findings

Chapter 8 compares the advantages and risks associated with each of the options discussed in the earlier chapters. A decision-tree approach is described that recommends various approaches, taking into consideration the nature, condition, and use of the records.

REFERENCES

Calmes, A., K. R. Eberhardt, and K. Kabadar. 1982. Pilot survey by National Archives and Records Service (NARS) and National Bureau of Standards (NBS). Unpublished.

Calmes, A., R. Schofer, and K. R. Eberhardt. 1985. National Archives and Records Service (NARS) Twenty Year Preservation Plan. NBSIR 85-2999, U.S. Department of Commerce, Gaithersburg, Maryland.

National Archives and Records Service. 1982. Staff Information Paper 21, Intrinsic Value in Archival Material.

National Association of Government Archives and Records Administrators. 1986. Preservation Needs in State Archives. Albany, New York: National Association of Government Archives and Records Administrators.

Pasadena, California, under heavy smog conditions and on a clear day. Outdoor pollutants contribute significantly to the deterioration of historical records.

3

Environmental Criteria

Long-term preservation of the paper-based collections at the National Archives requires that damage caused by environmental conditions and atmospheric pollutants be prevented. Because these documents, or at least their information content, must be retained indefinitely, even very slow rates of deterioration caused by air pollutants could lead to unacceptable levels of accumulated damage over a period of several hundred years. The problem of protecting the National Archives inventory is thus quite different from the question of protecting common consumer products from premature deterioration over their short service lifetime. Standards adopted for acceptable air quality outdoors therefore are not applicable, and separate air quality objectives must be set that are suited to the problem of long-term preservation of archived materials.

MATERIALS CONSIDERED

At the outset, it is necessary to recognize that document collections contain much more than just paper. Paper-based records sometimes are written in colored inks. Maps may be printed with colored inks or tinted with pigments. Some records are bound into volumes, and those bindings may contain thread, cardboard, adhesives, cloth, leather, and synthetic or chemically impregnated fabrics. Air pollution can cause damage to all these materials. In addition, paper-based records are sometimes converted to other media (e.g., photographic film or magnetic tape), and those materials also must then be protected.

AIR POLLUTANTS

Pollutants present in outdoor air may be drawn into buildings by conventional ventilation systems. Under authority derived from the Clean Air Act (42 USC 1857 et seq.), the U.S. Environmental Protection Agency (EPA) has adopted National Ambient Air Quality Standards for six common outdoor pollutants:

TABLE 3-1 National Ambient Air Quality Standards

Pollutant	Averaging Time[a]	National Standard Primary[b]	Secondary[c]
Ozone	1 hour	235 μg/m^3 (0.12 ppm)	Same as primary standard
Carbon monoxide	8 hour	10 mg/m^3 (9 ppm)	Same as primary standard
	1 hour	40 mg/m^3 (35 ppm)	
Nitrogen dioxide	Annual average	100 μg/m^3 (0.05 ppm)	Same as primary standard
Sulfur dioxide	Annual average	80 μg/m^3 (0.03 ppm)	—
	24 hour	365 μg/m^3 (0.14 ppm)	—
	3 hour	—	1300 μg/m^3 (0.5 ppm)
Suspended particulate matter	Annual geometric mean	75 μg/m^3	60 μg/m^3
	24 hour	260 μg/m^3	150 μg/m^3
Lead	Calendar quarter	1.5 μg/m^3	Same as primary standard

[a]Averaging times shown are the durations over which measurements are averaged for comparison to the standards.
[b]Primary standards are set to protect human health.
[c]Secondary standards that differ from primary standards are set in response to welfare effects including damage to materials.

SOURCE: Environmental Protection Agency (1971, 1978b, 1979).

sulfur dioxide (SO_2), nitrogen dioxide (NO_2), ozone (O_3), carbon monoxide (CO), suspended particulate matter, and lead, as shown in Table 3-1. The primary standards shown are at levels deemed necessary to protect human health, while the secondary standards for SO_2 and particulate matter have been adopted to take additional steps to slow the rate of damage to the public welfare (e.g., plant life, animal life, and materials) from those pollutants. Because of their regulated status and because they have been readily measurable for many years, much literature has been accumulated that can be used to characterize the atmospheric loading and the damage potential of these pollutants (Environmental Protection Agency, 1978a, 1982a, 1982b).

A variety of additional air pollutants are recognized for which outdoor air quality standards have not been set at a national level. A selected list of unregulated pollutants is shown in Table 3-2. These unregulated pollutants include acid gases [nitric acid (HNO_3), nitrous acid (HONO), formic acid (HCOOH), acetic acid (CH_3COOH), hydrochloric acid (HCl)], oxidants [peroxyacetyl nitrate ($CH_3COO_2NO_2$) and hydrogen peroxide (H_2O_2)], and reduced nitrogen and sulfur compounds (NH_3, H_2S) that fall within categories known to cause damage to materials such as those found in archives.

In the case of particulate air quality, difficulty arises if one tries to relate gross measures, such as total suspended aerosol mass concentration, directly to materials damage effects. A variety of chemically and physically distinct subfractions of the particulate matter complex can be identified that may have a higher potential for material damage than the bulk of the aerosol mass, including acid particles (e.g., H_2SO_4 mist), alkaline particles (e.g., cement dust), and black soot particles. Acid mists and soot particles are often concentrated in particle sizes less than 1 μm in diameter that are relatively difficult to remove when compared to coarser particles several micrometers in diameter. Unless a more sophisticated definition of particulate air quality is adopted than one based solely on total aerosol mass concentration, there is a danger that ventilation systems will be designed that will lower mass loadings without achieving a proportionate reduction in damage potential.

In addition to pollutants commonly found outdoors, air quality in buildings can be affected by contaminants emitted indoors. Indoor generation of pollutants in museums, archives, and libraries recently has been reviewed by Baer and Banks (1985a). They report a variety of sources, including formaldehyde and alkaline

TABLE 3-2 Partial List of Unregulated Gaseous Contaminants Observed or Possibly Present in Polluted Outdoor Air

Compound	Typical (or Maximal) Concentration Reported, ppm
1. Compounds observed in photochemical smog[a]	
Peroxyacetyl nitrate, $CH_3COO_2NO_2$	0.004 (0.01)
Hydrogen peroxide, H_2O_2	(0.18)
Formaldehyde, CH_2O	0.04
Higher aldehydes, RCHO	0.04
Acrolein, CH_2CHCHO	0.007
Formic acid, HCOOH	(0.05)
2. Compounds that may be formed in photochemical smog[a,b]	
Peroxybenzoyl nitrate, $C_6H_5COO_2NO_2$ (PB_zN)	
Nitric acid, $HONO_2$	
Organic hydroperoxides, ROOH	
Organic peracids, $RCOO_2H$	
Organic peroxynitrates, RO_2NO_2	
Ozonides, O_3-olefin	
Ketene, CH_2CO	
Nitrous acid, HONO	
Pernitric acid, HO_2NO_2	
Pernitrous acid, HO_2NO	
Sulfoxyperoxy nitrate, $HOSO_2O_2NO_2$	
3. Other unregulated air pollutants	
NH_3, H_2S, HCl	

NOTE: A comprehensive 24-page list of more than 150 chemical substances that are regulated as air pollutants in jurisdictions throughout the world is given by Newill (1977).

[a]Committee on Medical and Biologic Effects of Environmental Pollutants (1977).
[b]Some of the compounds listed under number 2 are speculative; concentrations have not been quantified.

Historical records requiring immediate preservation to avoid loss of information. Damage can result from deterioration of the paper, ink, and binding as well as from handling.

particles (setting concrete) released from building materials; corrosion inhibitors (e.g., diethylaminoethanol) introduced from improperly designed air humidification systems; oxides of nitrogen generated by decomposition of cellulose nitrate found in some photographic film, "acetate" recording disks, adhesives, and pyroxylin-coated or -impregnated fabric (often used in library rebindings); and organic acids (e.g., formic, acetic, and tannic acid) that are released by off-gassing from certain wood products and from decomposition of adhesives (e.g., polyvinyl acetate).

OBSERVED DAMAGE

The following discussions cover previously observed damage to materials similar to those found in archives.

Paper

Chapter 4 of this report indicates that acidification of paper leads to destruction of its mechanical properties via hydrolysis of the cellulose of the paper. Absorption of acid gases by paper can accelerate this acidification process, with resulting increased hydrolysis. Sulfur dioxide is readily absorbed by uncoated wallpaper, but the absorption process can be retarded by vinyl coatings on the paper surface (Spedding and Rowlands, 1970; Walsh et al., 1977). Once absorbed by the paper, SO_2 can be oxidized, thereby contributing to the acidity of the paper. Examination of book collections has shown that acidity is highest at the exposed outer edges and declines toward the center of the pages, a phenomenon that has been associated with SO_2 absorption at the exposed edges of the books (Parker, 1955; Hudson, 1967). Accelerated hydrolysis of paper by acid gases other than SO_2 is less well documented. As will be discussed shortly, reduction in the strength of cotton textile fibers has been observed from exposure to ambient air containing NO_2 in the presence of sunlight. It is reasonable, therefore, to expect that NO_2 and other acid gases may promote the failure of cellulose fibers in paper.

Paper-based materials are subject to deterioration by oxidation as well as by hydrolysis. Ozone will react with cellulose. However, its effect at ambient concentrations on the storage properties of paper records is not well documented. Although dose-response relationships for pollutant damage at low concentrations have not been experimentally established, one can infer a linear relationship based on measurements made at the ppm level (H. H. G. Jellinek, presentation to the committee, 1985).

Soiling of paper can result from the deposition of atmospheric particulate matter. Perceptual experiments show that white paper should appear soiled when only 0.2 percent of its surface has been covered with black deposited particles (Carey, 1959). Experiments by Hancock et al. (1976) confirmed that with maximum contrast a 0.2 percent effective area coverage by black particles represents the median threshold for detection of soiling by human observers. Hancock and his coworkers also coated common household items (including bond paper) with charcoal aerosol deposits at a variety of surface coverage densities. The median response of a panel of human subjects indicated that the test articles were "unfit for use" once the effective area coverage by black particles reached 0.7 percent. The relationship between atmospheric particulate matter loading and soiling of surfaces (e.g., paper) is not completely understood. "The poorly understood deposition rates and poorly characterized chemical and physical properties related to reflectance make general application of . . . [damage] functions difficult if not impossible" (Environmental Protection Agency, 1982a). However, important insights into the likely nature of the soiling of paper can be inferred from studies of the optical and chemical properties of ambient particulate matter collected on filters.

Recent atmospheric optical studies show that light absorption in urban atmospheres is dominated by the presence of small amounts of black carbonaceous material having a structure similar to that of impure graphite, often referred to as graphitic carbon, elemental carbon, or sometimes just soot. Examination of the decrease in reflectance of paper on which atmospheric particles have been collected by filtration shows that the reflectance decrease is due principally to the elemental carbon content of the aerosol (Cass et al., 1984). Diesel engine soot is a prominent source of elemental carbon in cities, although more than 50 other sources can be identified that contribute carbon particles to the atmosphere (Cass et al., 1984). These elemental carbon particles are found predominantly in fine particle sizes, below 2 μm in diameter, and represent only a very small fraction of the ambient aerosol mass (about 5.5 μg/m^3 annual mean at downtown Los Angeles in 1975, versus a total aerosol mass of over 100 μg/m^3 in that year at that site) (Cass et al., 1984). The characteristic color of coarse particle samples (diameter greater than 2 μm) is brown because of the presence of soil or road dust. Thus, the appearance of deposited particles would be expected to vary as a function of size and chemical composition.

While the soiling of paper as a function of particle size and chemical composition has not been extensively studied, the chemical composition of soiling deposits on Plexiglas display cases in a Los Angeles museum has been examined (Druzik, personal communication, 1984). These deposits appeared on the unwiped inside surfaces of the display cases over only a 3–month period following the opening of an exhibition. Chemical analysis showed that the deposits contained carbonaceous material, with 21 percent to 30 percent of the carbon present as black elemental carbon, which is the ratio of elemental carbon to total carbon close to that observed in atmospheric aerosol samples in downtown Los Angeles (Cass et al., 1984). This study, plus experience with particle samples filtered from the atmosphere onto paper substrates, suggests that *both* coarse and fine particles must be controlled if these colored atmospheric materials (i.e., elemental carbon and soil dust types) are to be eliminated as a potential source of soiling.

Leather

Leather is a common material found in the bindings of old books and documents. Absorption of SO$_2$ by leather is rapid, resulting in hydrolysis of the leather material, followed by cracking and eventual powdering of the leather (Spedding et al., 1971; Yokom and Grappone, 1976).

Textiles

Cloth and thread are used in library and document collection bindings. Cellulosic fabrics, like cotton, rayon, and certain types of nylon, are particularly suscep-

Bound volumes in storage stacks at the National Archives. Leather and cloth, as well as paper, pigments, and inks, are affected by environmental conditions.

tible to air pollutant damage. Upon exposure to SO_2, the breaking strength of cotton is reduced (Brysson et al., 1967; Zeronian, 1970; Zeronian et al., 1971). The strength of cotton also was reduced by exposure to sunlight and ambient air in Berkeley, California, under circumstances that implicate NO_x species as the damaging agents (Morris, 1966). Like paper, cellulosic fabrics can degrade by oxidation. Studies of the effect of ozone exposure on cotton textiles (Bogarty et al., 1952; Morris, 1966) showed a loss in tensile strength in the wetted samples studied, but no apparent loss in dry fabric samples. Textiles used in bookbinding would be expected to be vulnerable to soiling by atmospheric particulate matter (see previous discussion of soiling hazard to paper).

Dyes, Pigments, and Inks

In the late 1930s it was found that commercial textiles treated with a blue anthraquinone dye reddened when exposed to nitrogen dioxide (Rowe and Chamberlain, 1937; Salvin et al., 1952). NO_2-resistant dyes were formulated and used in comparative tests during the mid-1950s in several cities believed to represent different levels of pollutant exposure. It was found that textiles exposed to the atmosphere in Ames, Iowa, still faded rapidly but without noticeable reddening and that this second type of fading was due to the reaction with atmospheric ozone (Salvin and Walker, 1955; Salvin, 1969). Recent studies (Shaver et al., 1983) show that several widely used artists' pigments will fade rapidly upon exposure to ozone at levels found in Los Angeles photochemical smog. Such ozone-fugitive pigments identified to date include the alizarin lakes and natural yellow pigments used in Japanese woodblock prints. Air pollution-induced fading of inks used in preparing written and printed documents has not been studied to date, but given the chemical similarities between inks and other types of colorants (dyes and pigments) one may assume that at least some inks are pollutant-sensitive.

Adhesives

Although only limited literature on pollution-induced failure of adhesive joints exists, substantial evidence can be found of pollution damage to the polymers used in formulating adhesives. From these data, one may expect oxidation in these polymer systems and acid hydrolysis in special cases (H. H. G. Jellinek, presentation to the committee, 1985).

Corrosion of Metals

In common experience, paper clips, staples, and other metal fasteners corrode and stain paper-based documents stored in poorly controlled environments. However, such effects have seldom been observed in documents stored under good environmental conditions at the National Archives, and virtually none for the post-1940 records.

Photographic Film

As is discussed in Chapter 5, photographic film is sensitive to NO_2 exposure (Carroll and Calhoun, 1955). Exposure to atmospheric oxidants (e.g., ozone or peroxides) can cause the formation of microblemishes (McCamy, 1964; Henn et al., 1965; Weyde, 1972). Since information often is recorded at high magnification on microfilm, there is concern about the development of even small defects such as microblemishes.

Unregulated Pollutants

In the foregoing summary, little mention was made of any damaging effects of unregulated pollutants like those listed in Table 3-2. Most of the listed pollutants simply have not been tested in combination with the materials of interest to see if damage will result. This information vacuum should not be misinterpreted as indicating that the unregulated pollutants have no effect.

INDOOR POLLUTANT LEVELS

Whether or not a particular air pollutant represents a threat to archived materials depends on whether or not it is found in the indoor atmosphere of buildings. Transfer of the criteria pollutants (e.g., SO_2, NO_2, O_3, and particulate matter) from outdoors to the indoor atmosphere of libraries, archives, and museums has been studied in a few cases. Data available for buildings in a variety of cities can be organized so that the percentage attenuation of outdoor pollutant levels on introduction to the indoor environment by building ventilation systems is apparent.

Sulfur Dioxide Levels

In the absence of deliberate pollutant removal, the indoor level of the pollutants generated outdoors might be expected to approach the level in the outdoor air that feeds building ventilation systems. Thomson (1965) reported that sulfur dioxide levels inside non-air-conditioned spaces at the National Gallery in London and at the Victoria and Albert Museum range from 50 percent to 100 percent of the outdoor concentration. Recent SO_2 measurements by Hackney (1984) in areas of the Victoria and Albert Museum, which lacks an SO_2 removal system, show SO_2 levels of 9 ppb in a typical internal gallery when the level outside is 22 ppb. Measurements at the National Archives Building in Washington in December 1982 and January 1983 show a grand average of 9 ppb of SO_2 inside the building versus 23 ppb outside at 24th and L Streets, N.W. (Hughes and Myers, 1983). The Archives Building at present is equipped for coarse particle filtration only, and Hughes and Myers (1983) concluded that SO_2 appears to pass through that air conditioning system with little or no change.

Nitrogen Oxides

Investigators have examined oxides of nitrogen levels in archives and galleries that lack chemically protected air conditioning systems. Hackney (1984) found that NO_2 levels inside the Tate Gallery in London were highest in unconditioned galleries (at values of 15 to 23 ppb) but were lower inside loosely fitting display cases and unused storerooms (at values of 2 to 3 ppb NO_2). This suggests that enclosing archived documents in proper storage containers may afford significant protection from NO_2 exposure. NO_x levels inside the National Archives Building in Washington, measured by Hughes and Myers (1983), showed indoor NO_x concentrations in the range of 10 to 252 ppb compared with 10 to 527 ppb outside (the indoor tracking the outdoor). They concluded that the building ventilation system did not significantly attenuate outdoor NO_x levels. Indoor versus outdoor relationships for NO and NO_2 recently have been measured in the newly constructed Virginia Steele Scott Gallery at the Huntington Library in San Marino, California (Cass, presentation to the committee, 1985). Pollutant removal systems at that site are confined to particle filtration only. Over the period October 30–November 9, 1984, indoor NO averaged 37 ppb compared to 36 ppb outdoors, while indoor NO_2 averaged 38 ppb compared with 44 ppb outdoors. These studies suggest that indoor NO_x concentrations in well-ventilated galleries, libraries, and archives that lack a NO_x removal system will be expected to be close to those outdoors.

Ozone Levels

Ozone is a highly reactive gas and reportedly can be removed almost completely from the atmosphere of some museum galleries. Thomson (1978) noted that this is probably due to O_3 interaction with interior surfaces (presumably including the collections). Studies by Shair and Heitner (1974) showed that indoor ozone levels can be predicted by a simple mathematical model given ozone loss rate data for various building materials, data on building indoor surface characteristics, ventilation rates, and whether or not a deliberate pollutant removal system is present.

A survey of ozone levels inside museums and libraries recently has been completed in Southern California (Cass, presentation to the committee, 1985), from which useful generalizations can be drawn about the ozone level expected as a function of building design. Buildings with rapid air exchange with the outdoors, no internal air recirculation, hard interior surfaces, no ozone removal equipment, and a high volume-to-surface area ratio showed indoor O_3 levels 70 percent to 80 percent of the values found outside. Peak 1-hour average ozone levels inside one such museum in Los Angeles have been observed at 143 ppb compared with 173 ppb outside, or 83 percent of the outdoor level. This is consistent with observed values in the gallery at the Sainsbury Center for Visual Arts in England, where indoor O_3 levels of up to 40 ppb were observed inside a modern art gallery in the presence of peak outdoor levels of 58 ppb, or 69 percent of the outdoor level (Davies et al., 1984). Southern California galleries with conventional air conditioning systems, a high internal air recirculation rate, but no ozone removal equipment, showed indoor ozone levels about one-third that observed outdoors. Buildings without air conditioning that had little ventilation often showed very low O_3 levels (about 10 percent of that outdoors), again due to depletion by reaction with building surfaces.

Measurements made inside the National Gallery, the Madison Building of the Library of Congress, and the National Archives Building in winter by Hughes and Myers (1983) showed undetectably small O_3 levels. They attributed this to O_3 loss to interior surfaces and cautioned the reader that conditions during the summer high-ozone season may be quite different. Indoor O_3 levels in the Archives Building during the summer are unknown but should be investigated before making an assumption that they are very low.

Particulate Levels

Most libraries, archives, and galleries currently employ particle filtration equipment, and therefore particulate levels are reviewed in detail in the following discussion of pollutant removal systems.

POLLUTANT REMOVAL SYSTEMS

Sulfur Dioxide Removal

Sulfur dioxide removal from building ventilation air has been achieved successfully by a variety of means. Hughes and Myers (1983) showed that application of a wash system at the East Building of the National Gallery in Washington reduced SO_2 levels to below 1 ppb. At the Madison Building of the Library of Congress in Washington, a pollutant removal system based on a packed bed of Purafil ($KMnO_4$ on an alumina support) reportedly reduced SO_2 levels below 0.5 ppb (Hughes and Myers, 1983). Hackney (1984) examined SO_2 concentrations in the new extension galleries at the Tate Gallery in London, where the air conditioning system employs activated carbon filters. He found that SO_2 levels were reduced to 0 ppb compared with 26 ppb outside on February 4, 1980, and to between 4 ppb and less than 2 ppb compared with 80 ppb SO_2 outside on March 14, 1980.

Nitrogen Oxides Removal

The NO_x removal efficiency of acid gas control systems in actual use in the Washington, D.C., area also has been examined by Hughes and Myers (1983). At the East Building of the National Gallery, a wash system reduced the indoor NO_x levels to the range of 7 to 50 ppb during times when outdoor levels were in the range of 40 to 92 ppb. In the Madison Building of the Library of Congress, a packed bed of Purafil for acid gas removal reduced the indoor NO_x levels to the range of 4 to 154 ppb in the presence of outdoor levels of 46 to 318 ppb. These data show that the wash and Purafil systems are much less effective for NO_x removal than for SO_2 removal, but the reason is not yet clear, since laboratory tests show Purafil to be effective. Additional research is needed to identify appropriate NO_2 removal practices. These test data should be reviewed to ascertain whether any NO_2 measurements were made, as opposed to measurements of total NO_x. NO_2 and other species (e.g., HNO_3) that are measured as if they were NO_2 by chemiluminescent NO_x monitors are the damaging pollutants of interest. It may be that the Purafil and wash systems are removing much of the NO_2 but leaving NO uncollected. This would contribute to high indoor NO_x levels yet provide a low NO_2 level indoors. Therefore, a second set of indoor versus outdoor NO_2 measurements should be commissioned, if necessary, to check indoor NO_2 levels explicitly. In addition, the NO_2 removal efficiency of activated carbon or chemically impregnated activated carbon-based pollutant removal systems in actual use should be examined.

Ozone Removal

Activated carbon filtration systems for ozone removal are used by many major libraries and museums in the Los Angeles area. These include the Huntington Library, Huntington Library Art Gallery, Los Angeles County Museum, Norton Simon Museum, J. Paul Getty Museum, and Southwest Museum Library. Indoor versus outdoor ozone measurements made at the Huntington Library Art Gallery in midsummer of 1984 showed peak indoor O_3 levels of 10 ppb when peak outdoor levels were 170 ppb, or about 94 percent removal (Cass, presentation to the committee, 1985).

Particulate Matter Removal

Even though most forced ventilation systems contain some form of particle filtration device, very little literature exists on the detailed effect of these filters on the chemical and soiling aspects of air quality in archives, libraries, and museums. Particulate matter concentrations were not measured during the recent examination of the National Archives Building, the National Gallery, and the Library of Congress (Mathey et al., 1983). This lack of data could be perpetuated if the monitoring procedures outlined by Mathey et al. (1983) are adopted, because these procedures involve monitoring the pressure drop across the particle filters rather than their actual in-use particle removal performance as a function of aerosol size and composition. If there is no direct examination of the actual measure of particle breakthrough or indoor aerosol levels, then it is possible to miss the identification of those particles that are not being collected effectively, or, alternatively, to miss the fact that particulate matter is being generated indoors.

Particulate air quality control considerations applicable to museums and other places where irreplaceable materials are stored were discussed by Thomson (1965, 1978). A thorough assessment of particle removal systems must take particle size and chemical composition into account. The size distribution of atmospheric particulate matter usually is bimodal or trimodal, with a coarse particle grouping consisting largely of soil dust, road dust, and sea salt (particle diameters of about 1 or 2 μm and larger); an ultrafine particle mode consisting of freshly nucleated gas-to-particle conversion products, such as H_2SO_4 aerosol (in sizes below 0.1 μm); and an accumulation-mode aerosol (particle diameter ranging between 0.1 μm and 1 or 2 μm) derived from fresh emissions of combustion products or the coagulation of ultrafine aerosol plus condensation onto a preexisting aerosol. Under polluted urban conditions, the coarse and fine particles may make comparable contributions to total particle volume (and hence mass) concentration (see Figure 3-1). The fine particles, below 1 or 2 μm in diameter, are chemically quite different from coarse dust and may have unique damage potential because of their high black soot content and potential for contributing acid aerosols.

Indoor-outdoor particulate matter concentration relationships have been reviewed (Yokom et al., 1976; Meyer, 1983), and the authors noted that indoor levels can be greater than or less than those outdoors, depending on ventilation conditions and on activity levels inside the buildings. An examination was made of particulate matter concentrations inside and outside the City Hall, a non-air-

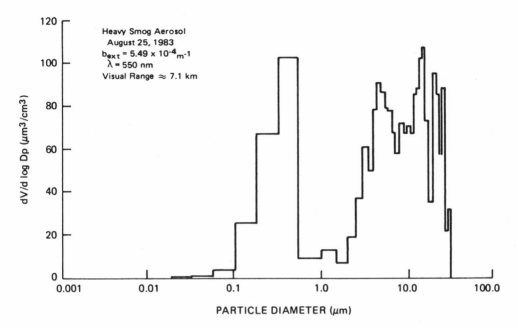

FIGURE 3-1 Size distribution of heavy smog aerosol at Pasadena, California (Larson et al., 1984).

conditioned library, two air-conditioned office buildings, and two private homes in Hartford, Connecticut (Yokom et al., 1971). Indoor particulate matter levels relative to those outdoors ranged from 0.16:1 to 1.15:1, with indoor levels less than those outside in all but one case. For the air-conditioned buildings, the indoor-to-outdoor aerosol mass ratio ranged from 0.31:1 to 0.75:1. The largest percentage attenuation of outdoor loadings occurred during winter episodes when outdoor mass concentrations were high. This observation was explained by noting that the high-concentration events in Hartford were enriched in large particles that are removed readily as they attempt to penetrate a building. Both aerosol mass concentration and the soiling index were monitored during the study, and a greater attenuation of the aerosol mass concentration than of the soiling index between outdoors and indoors was found. That observation is consistent with the hypothesis that soot particles contributing to the soiling index values are primarily concentrated in the fine particle sizes that may negotiate building air inlets more readily than coarse particle material. The assessment concluded that indoor-outdoor soiling index ratios in that study did not appear to be significantly different for air-conditioned versus non-air-conditioned buildings, which indicates that soiling particles were not effectively removed by the filters in the air-conditioning systems.

Further evidence of the selective alteration of the chemical composition and size of airborne particulate matter by ventilation systems is illustrated by data taken at a newly built art gallery in Southern California. The gallery's particle removal system specifications called for U.L. Class 2 filters, Farr 30/30 or equal, pleated, strainer mat-type, 2 in. thick. Aerosol mass loading was measured inside and outside of the gallery, and the chemical composition of the aerosol at both sites was analyzed for ionic species by ion chromatography, for trace elements by X-ray fluorescence and atomic absorption, and for organic and elemental carbon by temperature-programmed combustion and pyrolysis. Indoor aerosol concentrations of 30.9 $\mu g/m^3$ were found compared with 79 $\mu g/m^3$ in the outdoor air supplied to the air-conditioning system (Cass, presentation to the committee, 1985). Chemical analyses of the indoor and outdoor samples showed 90 to 100 percent removal of the crustal elements Al, Si, Ca, Ti, Mn, and Fe, indicating excellent removal of coarse particle soil dust.

Chemical elements characteristically found in fine particle sizes less than 2 μm in diameter were removed with very poor efficiency: Pb and Br customarily associated with automobile exhaust were removed only to a slight extent (19 percent removal and 9 percent removal, respectively). Between 16 percent and 55 percent of the sulfate aerosol was removed by the ventilation system. Total aerosol carbon particle levels inside were almost identical to total aerosol carbon levels outside, but black elemental carbon levels indoors were lower than those outdoors, suggesting an indoor source of aerosol organic carbon. This poor filtration efficiency for fine particles should be avoided in future designs, since the fine particle fraction of the outdoor aerosol burden contains much of the black soot and acidic material.

Fine particle control can be achieved by high-performance filters with or without simultaneous use of electrostatic precipitators. Thomson (1965) warned that electrostatic precipitators should not be used in the air-conditioning systems of museums because of the potential for ozone generation. The NBS study (Mathey et al., 1983) tended to discount this problem because the O_3 levels generated would be below their recommended indoor air quality standard. Thomson's (1965) caution is appropriate. The committee considers the NBS O_3 air quality limit to be too high, and experience shows that it is unwise to assume that a carbon bed used to protect against the deliberate generation of O_3 in an electrostatic precipitator will be working properly at all times.

RECOMMENDED STANDARDS

The critical areas identified for the National Archives indoor air quality control are covered individually in the following sections.

Temperature and Relative Humidity

The control of temperature and relative humidity is frequently cited as the first step in environmental control in collections management (Thomson, 1978; Mathey et al., 1983). Indeed, the evidence provided by the committee's visit to the Archives suggests that the resulting benefits are well documented in actual practice. However, precise temperature and relative humidity standards are less readily identified. The rationale for the lowest temperature of storage consistent with energy conservation (costs) and worker comfort lies in the Arrhenius relationship (reaction rate) and its consequence that a reduction in temperature of 10°C reduces the rate of reaction (e.g., oxidation, hydrolysis) by approximately a factor of 2. The committee is not aware of any data that support precise levels of control on either side of the selected temperature. Hence, the committee has specified a temperature range to be maintained, rather than a single temperature with an artificially precise level of control about that temperature. No benefits are known to be derived from controlling a temperature to $\pm 1°F$ ($0.5°C$) in contrast to $\pm 2°F$ ($1.1°C$) or even $\pm 5°F$ ($2.8°C$).

The specification of relative humidity (RH) is more difficult, since higher levels of RH generally are desirable for materials handling, whereas low RH is preferable for reduction of biological or chemical attack. (See Chapter 4 for a discussion of the effects of temperature and moisture variations on the properties of paper.) Further complications arise in mixed collections that involve leather, textiles, paper, and photographic materials. The prudent approach appears to involve the selection of a median RH level of approximately 40 to 50 percent with a modest level of control, since fluctuations in RH (more appropriately equilibrium moisture content) introduce undesired mechanical stresses. This is especially so in bound volumes, larger format paper documents, and photographic materials. It also should be noted that the sensors used for control in active systems (e.g., air conditioning humidity sensors) are notoriously unreliable, so that specified control is often ephemeral. Here, too, the literature provides little support for benefits associated with a more precise level of active control than ± 5 percent RH. The committee emphasizes its belief that the temperature and relative humidity values specified should be achieved at the document surface, not simply within the room air. This suggests greater reliance on the demonstrated buffering capacity of controlled microenvironments, such as polyester encapsulation and acid-free boxes, than on active control.

Air Quality

The committee feels that sufficient information exists to demonstrate that materials like those found in library collections can be damaged if stored in poor environmental conditions. At the same time, the experimental data needed to quantify dose-response functions for use in making precise damage predictions do not exist. For that reason, recommended environmental conditions for storage of archived materials at present are based primarily on expert opinion. Tables 3-3 and 3-4 summarize the variety of recommendations that have been made.

Evidence that good environmental control will make a significant difference to the future of the National Archives is provided by a direct inspection of the present Archives collection by members of the committee who are experts on the condition of the Library of Congress collection. The Archives first installed its conventional air conditioning in the 1930s, while the Library of Congress did not achieve temperature and humidity control in most parts of its collection until the 1960s. The condition of the paper in bound volumes in the Archives at present was found to be noticeably better than that at the Library of Congress. Furthermore, the vast majority of the Archives collection traditionally has been housed in file boxes rather than in bound volumes. The committee's inspection team reports that the paper records within these boxes are in better condition than the paper records in bound volumes on the same shelves. Bound materials typically show evidence of deterioration and aging that progresses into the volumes from the outer edges of the paper, which are exposed to room air. The paper in boxed

TABLE 3-3 Air Quality Criteria for Archives, Libraries, and Museums

Authority or Installation	SO_x	NO_x	O_3	Particulate Removal Required
ANSI-PH	Suitable washers or absorbers			Preferably HEPA
ASHRAE	Canister-type filters or spray washers of chemical pollutants in outdoor air			85% DSM
BML	0	0	0	0
CCI	Should not exceed 10 ppb			95% ≥ 1 μm
	Consider central air purification in high ambient areas			50% $0.5-1$ μm
LC	Purafil system in use			95%
NBS	1 μg/m^3 (0.4 ppb)	5 μg/m^3 (2.5 ppb NO_2)	25 μg/m^3 (13 ppb)	75 μg/m^3 TSP (HiVol)
N-PNB	≤ 10 μg/m^3	≤ 10 μg/m^3	≤ 2 μg/m^3	High-rating DSM
ROM-C	Charcoal or equivalent filtration to remove SO_x, NO_x, O_3			99% ≥ 10 μm / 95% ≥ 1 μm
T	≤ 10 μg/m^3	≤ 10 μg/m^3	0–2 μg/m^3	60–80% MBT
ANSI-DSP	≤ 1 μg/m^3 (0.4 ppb)	N.S.	≤ 2 μg/m^3 (1 ppb)	See Table 3-4

KEY: ANSI-PH = American National Standards Institute—Photographic Standards; ASHRAE = American Society of Heating, Refrigeration, and Air Conditioning Engineers; BML = British Museum Libraries; CCI = Canadian Conservation Institute; LC = Library of Congress (Madison Building); NBS = National Bureau of Standards; N-PNB = Newberry Library—PN Banks Planning Study; ROM-C = Royal Ontario Museum Conference; T = G. Thomson; ANSI-DSP = American National Standards Institute Practice for Storage of Paper-Based Library and Archival Materials (Draft 4, 1985); HEPA = High-Efficiency Particulate Air; DSM = Dust Spot Method; TSP = Total Suspended Particulates; MBT = Methylene Blue Test; N.S. = No Standard.

SOURCE: After Baer and Banks (1985a).

containers shows minor mechanical damage, mainly at the top edges of files, and this damage is probably due to repeated handling during searches. The microenvironment within the boxes appears to protect the records. This is consistent with (a) the damping of temperature and humidity fluctuations by the box and (b) the presence of a barrier against pollutant intrusion.

This evidence, plus the realization that the vast majority of the Archives collection will remain on paper for the foreseeable future, argues in favor of the committee's endorsement of both the draft ANSI standard, Practice for Storage of Paper-Based Library and Archival Materials, and the suggested NBS standards for proposed environmental control in archives. Where the two sets of recommendations are in conflict, the more restrictive requirement is endorsed (e.g., the proposed ANSI recommendations for O_3 and particulate matter). The NBS recommendations for particulate matter are derived from ASHRAE recommendations, which are based largely on human occupancy requirements. These requirements could be met without providing any significant protection from much of the fine black particulate matter that would cause a long-term soiling hazard to the Archives. On the other hand, the ANSI standards reflect the committee's concern that these fine particles are detrimental and should be removed. At present, the majority of the experts cited in Tables 3-3 and 3-4 view ozone as being at least as hazardous to many materials as is NO_x. Therefore, it can be concluded that the concentration objectives for the gaseous pollutants (e.g., SO_2, NO_2, O_3) should be

TABLE 3-4 Draft ANSI Particulate Standards for Paper-Based Documents in Libraries and Archives

System Filter Location	ASHRAE Weight Arrestance Efficiency	ASHRAE Atmospheric Dust Spot Efficiency	MIL-STD 282 DOP Efficiency
Prefilter[a]	≥80%	≥30%	≥5%
Intermediate filter[b]	≥95%	≥80%	≥50%
Fine filter[b]	N.A.	≥90%	≥75%

[a]For outside or makeup air.
[b]For supply (both outside and recirculated) air.

KEY: DOP = Dioctyl phthalate; N.A. = Not applicable.

SOURCE: After Baer and Banks (1985b).

controlled to the same general order of magnitude. The committee's suggested standards are given in Table 3-5.

For the Archives inventory, these environmental conditions need only be achieved at the surface of the documents. The Archives file boxes provide a micro-environment that probably helps to damp temperature and relative humidity fluctuations and also probably attenuates pollutant intrusion into the box. A study is needed of the transfer coefficients that relate ambient conditions in the Archives stack areas to conditions inside the boxes. Much of the intended environmental protection apparently can be provided by passive control (e.g., the boxes) rather than by complete reliance on expensive active control measures (e.g., air conditioning).

On the average, an individual page within the Archives collection is likely to be retrieved, consulted, or made use of less than once every 100 years. In view of this, the cost advantages of passive environmental control could be extended by placing much of the Archives collection in remote low-temperature storage (e.g., underground vaults) outside the city of Washington. These vaults could be

TABLE 3-5 Recommended Standards for Paper-Based Records in a Mixed Collection of Bound and Unbound Materials (Standards to be Met at the Surface of the Records)

Environmental Variable[a]	Control Level
Temperature	68–72°F
Relative humidity	40–50%
SO_2	≤1 $\mu g/m^3$ (0.4 ppb)
NO_2, HNO_3	Best available technology
O_3	≤2 $\mu g/m^3$ (1 ppb)
Particulates	Same as Table 3-4

[a]Specifications are averages over a 24-hour period. Small, short-term excursions outside these limits are permitted.

selected so that they present a lower air-conditioning and pollutant removal load than that required at the Archives Building, which is located in the city center.

Monitoring for Indoor Air Pollutant Objective Compliance

For the gaseous pollutants, SO_2, O_3, and NO_2, ambient measurements can be taken either by manual methods, in which an integrated sample is collected over a period of hours or days, or by use of continuous-monitoring instruments. Manual methods usually involve drawing an air sample through an appropriate liquid reagent, followed by colorimetric determination in the laboratory. Standard manual methods for SO_2, NO_2, oxidants, and many other gaseous pollutants are described by the Intersociety Committee (1977).

Continuous-monitoring instruments for SO_2, O_3, and NO_2 are customarily used for monitoring pollutant levels in outdoor air. Routine ambient monitoring systems in the Los Angeles area employ pulsed fluorescent SO_2 monitors, chemi-luminescent $NO/NO_2/NO_x$ monitors, and ultraviolet photometric O_3 monitors. These instruments operate without consuming wet chemical reagents and specialized gases. Instruments that operate by other equivalent methods are available.

A potential difficulty with most of the standard manual methods and continuous instruments on the commercial market today is that these systems were not designed to measure pollutants at the very low levels specified by the proposed ANSI or NBS objectives described earlier. The lower detection limit of the standard manual methods (typically 5 to 10 ppb) might be reduced by drawing larger volumes of air through the absorbing reagent, but this should be done only if the absorption efficiency of the system at that altered flow condition is confirmed.

Review of specification sheets supplied by several manufacturers of continuous-monitoring instruments shows detection limits as listed in Table 3-6. It will be noted that the minimum detection limits listed for O_3 and SO_2 are slightly higher than the proposed ANSI standards. These detection limits appear to be close enough to the stated objectives that a rational approach might well be first to design the building air conditioning system to meet the air quality objectives and then to monitor for equipment failure by determining whether pollutant levels exceed instrumental detection levels similar to those given in Table 3-6. If research into the effectiveness of storage boxes shows that pollutant levels in room air can be increased, then the minimum detection limits of present monitoring systems may cease to be a concern.

The proposed ANSI standards for particulate matter filtration are based on current ASHRAE test methods that can be used to determine that new filter media are performing as expected. Monitoring the system pressure drop should be sufficient to discover clogged filters that need to be replaced. In addition, particulate matter concentration measurements should be made in the archival storage areas, thereby guarding against the presence of unexpected indoor sources of particulate matter. Given the stringent filtration conditions in the proposed ANSI standards, indoor particle levels would be expected to be very low. If monitoring shows that this is not so, then the origin of the unexpected aerosol material should be investigated. Chemical analysis of the collected filter samples often can be used to identify the likely source of airborne particulate matter.

TABLE 3-6 Detection Limits of Continuous-
Monitoring Instruments

Pollutant	Measurement Method	Minimum Detection Limit
O_3	Ultraviolet photometric	2 to 3 ppb
SO_2	Pulsed fluorescent	2 ppb
NO_2	Chemiluminescent	2.5 ppb

In addition to selection of measurement methods, a decision must be made either to monitor building pollutant removal system performance continuously or to check indoor air quality at periodic intervals. The continuous-monitoring approach probably would require having a skilled air monitoring technician on the National Archives staff to maintain the equipment. Alternatively, the monitoring equipment might be maintained by arrangement with other local government agencies that currently operate continuous ambient monitoring systems for outdoor air quality. If staffing a continuous-monitoring system proves to be impractical, then an outside consulting firm might be employed. Its task would be to survey both indoor pollutant levels and the condition of absorbent materials extracted from pollutant removal system beds at periodic intervals (quarterly or semi-annually), thereby determining when the pollutant removal system materials must be renewed.

Other Considerations

Damage to organic matter (paper, film, tape, etc.) from rodents, termites, fungus, and bacteria has been effectively controlled by limiting access to the storage areas and maintaining effective environmental controls in these areas.

Damage from exposure to light is not a factor in the case of documents without intrinsic value. In most libraries, there is a minimum of direct sunlight, and artificial light has very little, if any, of the wavelengths in the ultraviolet (UV) range that are detrimental to organic inks and dyes (H. H. G. Jellinek, presentation to the committee, 1985). Furthermore, these documents are not customarily placed on public display, so their exposure to any kind of light is likely to be minimal. Light exposure does present something of a problem for records having intrinsic value, which are outside the purview of this study.

REFERENCES

Baer, N. S., and P. N. Banks. 1985a. Indoor air pollution: Effects on cultural and historic materials. Int. J. Museum Manage. Curatorship, 4:9–20.

Baer, N. S., and P. N. Banks. 1985b. Particulate standards for museums, libraries, and archives. 78th Ann. Meet. Air Pollut. Control Assoc. Preprint 85–8.8.

Bogarty, H., K. S. Campbell, and W. D. Appel. 1952. The oxidation of cellulose by ozone in small concentrations. Text. Res. J., 22:81–83.

Brysson, R. S., B. J. Trask, J. B. Upham, and S. G. Booras. 1967. The effects of air pollution on exposed cotton fabrics. J. Air Pollut. Control Assoc., 17:294–298.

Carey, W. F. 1959. Atmospheric deposits in Britain—A study of dinginess. Int. J. Air Pollut., 2:1–26.

Carroll, J. F., and J. M. Calhoun. 1955. Effect of nitrogen oxide gases on processed acetate film. J. Soc. Motion Pict. Telev. Eng., 64:501–507.

Cass, G. R., M. H. Conklin, J. J. Shah, J. J. Huntzicker, and E. S. Macias. 1984. Elemental carbon concentrations: Estimation of an historical data base. Atmos. Environ., 18:153–162.

Committee on Medical and Biologic Effects of Environmental Pollutants. 1977. Ozone and Other Photochemical Oxidants. Washington, D.C.: National Academy of Sciences.

Davies, T. D., B. Ramer, G. Kaspyzok, and A. C. Delany. 1984. Indoor/outdoor ozone concentrations at a contemporary art gallery. J. Air Pollut. Control Assoc., 31:135–137.

Environmental Protection Agency. 1971. Primary and Secondary Air Quality Standards. Fed. Reg. 36:22388–22392.

Environmental Protection Agency. 1978a. Air Quality Criteria for Ozone and Other Photochemical Oxidants. EPA-600/8-78-004. Research Triangle Park, North Carolina: U.S. Environmental Protection Agency.

Environmental Protection Agency. 1978b. National Air Quality Standard for Lead. Fed. Reg. 36:46245–46277.

Environmental Protection Agency. 1979. Revisions to the National Ambient Air Quality Standard for Photochemical Oxidants. Fed. Reg. 44:8201–8233.

Environmental Protection Agency. 1982a. Air Quality Criteria for Particulate Matter and Sulfur Oxides, Vol. III. EPA-600/8-82-029cF. Research Triangle Park, North Carolina: U.S. Environmental Protection Agency.

Environmental Protection Agency. 1982b. Air Quality Criteria for Oxides of Nitrogen. EPA-600/8-82-026. Research Triangle Park, North Carolina: U.S. Environmental Protection Agency.

Hackney, S. 1984. The distribution of gaseous air pollution within museums. Stud. Conserv., 29:105–116.

Hancock, R. P., N. A. Esmen, and C. P. Furber. 1976. Visual response to dustiness. J. Air Pollut. Control Assoc., 26:54–57.

Henn, R. W., D. G. Wiest, and B. D. Mack. 1965. Microscopic spots in processed microfilm: The effect of iodine. Photog. Sci. Eng., 9:3, 167–173.

Hudson, F. L. 1967. Acidity of 17th and 18th century books in two libraries. Pap. Technol., 8:189–190.

Hughes, E. E., and R. Myers. 1983. Measurement of the Concentration of Sulphur Dioxide, Nitrogen Oxides, and Ozone in the National Archives Building. NBSIR 83-2767. Washington, D.C.: National Bureau of Standards.

Intersociety Committee on Methods of Air Sampling and Analysis. 1977. Methods of Air Sampling and Analysis, Second Edition, M. Katz, ed. Washington, D.C.: American Public Health Association.

Larson, S. M., G. R. Cass, K. J. Hussey, and F. Luce. 1984. Visibility Model Verification by Image Processing Techniques. Environmental Quality Laboratory Report. Pasadena, California: California Institute of Technology.

Mathey, R. G., T. K. Faison, and S. Silberstein. 1983. Air Quality Criteria for Storage of Paper-Based Archival Records. NBSIR 83-2795. Washington, D.C.: National Bureau of Standards.

McCamy, C. S. 1964. Inspection of Processed Photographic Record Films for Aging Blemishes. National Bureau of Standards Handbook 96. Washington, D.C.: National Bureau of Standards.

Meyer, B. 1983. Indoor Air Quality. Reading, Massachusetts: Addison-Wesley.

Morris, M. A. 1966. Effect of Weathering on Cotton Fabrics. Bulletin 823. Davis, California: California Agricultural Experiment Station.

Newill, V. A. 1977. Air quality standards. Air Pollution, Third Edition, Vol. V, A. C. Stern, ed. New York: Academic Press.

Parker, A. 1955. The destructive effects of air pollution on materials. Proceedings of the 22nd Annual Conference, National Smoke Abatement Society, Bournemouth, England, September 28, 1955. Brighton, England: National Smoke Abatement Society.

Rowe, F. M., and K. A. Chamberlain. 1937. Fading of dyes on cellulose acetate rayon. J. Soc. Dyers Colour., 53:268–278.

Salvin, V. S. 1969. Ozone fading of dyes. Text. Chem. Color., 1:245–251.

Salvin, V. S., and R. A. Walker. 1955. Service fading of disperse dyestuffs by chemical agents other than the oxides of nitrogen. Text. Res. J., 25:571–585.

Salvin, V. S., W. D. Paist, and W. J. Myles. 1952. Advances in theoretical and practical studies of gas fading. Am. Dyest. Rep., 41:297–302.

Shair, F. H., and K. L. Heitner. 1974. Theoretical model for relating indoor pollutant concentrations to those outside. Environ. Sci. Technol., 8:444–451.

Shaver, C. L., G. R. Cass, and J. R. Druzik. 1983. Ozone and the deterioration of works of art. Environ. Sci. Technol., 17:748–752.

Spedding, D. J., and R. P. Rowlands. 1970. Sorption of sulphur dioxide by indoor surfaces. I: Wallpaper. J. Appl. Chem., 20:143–146.

Spedding, D. J., R. P. Rowlands, and J. E. Taylor. 1971. Sorption of sulphur dioxide by indoor surfaces. III: Leather. J. Appl. Chem. Biotechnol., 21:68–70.

Thomson, G. 1965. Air pollution—A review for conservation chemists. Stud. Conserv., 10:147–167.

Thomson, G. 1978. The Museum Environment. London: Butterworths.

Walsh, M., A. Black, A. Morgan, and G. Crashaw. 1977. Sorption of SO_2 by indoor surfaces including carpets, wallpaper, and paint. Atmos. Environ., 11:1107–1111.

Weyde, E. 1972. A simple test to identify gases which destroy silver images. Photog. Sci. Eng., 16:4, 283–286.

Yokom, J. E., and G. F. Grappone. 1976. Effects of Power Plant Emissions on Materials. EPRI EC-139. Palo Alto, California: Electric Power Research Institute.

Yokom, J. E., W. L. Clink, and W. A. Cote. 1971. Indoor/outdoor air quality relationships. J. Air Pollut. Control Assoc., 21:251–259.

Yokom, J. E., W. A. Cote, and F. B. Benson. 1976. Effects on indoor air quality. Air Pollution, Third Edition, Vol. II, A. C. Stern, ed. New York: Academic Press.

Zeronian, S. H. 1970. Reactions of cellulosic fabrics to air contaminated with sulfur dioxide. Text. Res. J., 40:695–698.

Zeronian, S. H., K. W. Alger, and S. T. Omaye. 1971. Reactions of fabrics made from synthetic fibers to air contaminated with nitrogen dioxide, ozone or sulfur dioxide. Proceedings of the Second International Clean Air Congress, H. M. Englund and W. T. Beery, eds. New York: Academic Press.

Record storage stacks showing older style of government filing system that required documents to be folded. The National Archives is currently refiling these records in covered box storage.

4

Paper

The word paper comes from papyrus, a sheet made by pressing together very thin strips of the Egyptian reed *Cyperus papyrus* (McGovern, 1978). However, papyrus sheets are not considered paper because the individual vegetable fibers are not separated and then reformed. Paper, by its traditional definition, must be made from natural fiber that has been macerated until each individual filament is a separate unit, the fibers are dispersed in water, and by use of a sieve-like screen the water is drained from the fiber, leaving a sheet of matted fiber on the surface of the screen. When dried, this thin layer of intertwined fiber is paper (Hunter, 1978). Modern paper manufacturing equipment employs this same principle in forming the paper web.

RAW MATERIALS AND STRUCTURE

Cellulosic plant-derived fibers are the raw materials that make up the major part of all papers. Natural plant fibers consist of crystalline filamentous cellulose that is the structure of the skeleton of the fiber. Chemically, cellulose is a linear polymer of beta-D-glucopyranose units linked by 1,4 glycosidic bonds. Isolated samples are found to have molecular weight varying from perhaps 50,000 to more than 1 million for a degree of polymerization of upwards of 7,000 and a length exceeding 0.003 cm. Cellulose has a monoclinic crystal structure characterized by a repeat distance of 1.03 nm (two anhydroglucose units) in the chain, with the repeat units assuming a chair configuration (Mark, 1983). X-ray evidence indicates that purified wood and cotton cellulose is about 70 percent crystalline. Lateral hydrogen bonds stabilize the crystal against relative displacement of the chains in response to imposed physical forces. Cellulose is a white substance that is hygroscopic in nature, insoluble in most solvents, and resistant to the action of most chemicals except strong acids. It is a stable organic polymer, and under suitable storage conditions it can be preserved for centuries or millenia without severe deterioration. Natural cellulosic fibers are structurally quite similar, and the fibers of cereal straws, bagasse, kenaf, bamboo, esparto, hemp, jute, flax, cotton, bark, and wood are used to manufacture paper. Wood pulp has proved to be the most important source of papermaking fiber (Emerton, 1980a, 1980b). Bleached chemical wood pulp and cotton fibers can be used to produce high-quality, stable papers.

The wall structure of cotton fibers is similar to that of wood fibers; it is relatively thin and grows free of lignin. The molecular weight of cotton cellulose at a degree of polymerization of 8,000 is slightly higher than that in wood, and the crystallites are slightly longer. The longer virgin fibers used in textiles are too valuable to be economical for paper (Rollins, 1965). However, the cotton ginning operation leaves a fuzz of short fibers on the cotton seed, and these shorter hair fibers, or linters, together with rags and textile clippings, are the sources of cotton fiber for special papers.

Kraft or sulfate pulp is the major wood pulp produced, and it is used for many grades of paper. If white paper or high-brightness pulp is to be produced, lignin and the hemicelluloses must be removed from the pulp fiber by bleaching. Multistage bleaching with agents such as chlorine, chlorine dioxide, caustic extraction, and peroxides is used to produce high-brightness pulp. Chemical wood pulps are classified according to the pulping process used—e.g., soda pulp, sulfite pulp, and kraft or sulfate pulp. These three chemical pulps in the fully bleached form are suitable for producing archival papers.

The remaining classes of commercial pulps—e.g., groundwood, semichemical, and thermomechanical—are not suitable for use in archival papers because of their lignin content. Unlike the highly stable crystalline cellulose, lignin is an amorphous, complex, polydisperse polymer network of phenylpropane units with a number of reactive functional groups that changes to a more highly colored form as it ages. For this reason, papers made with lignin-containing fibers tend to discolor with age (Sjostrom, 1981).

Wood fibers are typically from 1.0 to 5.0 mm in length and from 25 to 50 μm in width and about 5.0 μm thick. Because of their unique structure, wood fibers exhibit a higher strength-to-weight ratio than any other structural material, with a modulus of elasticity or Young's modulus of 3×10^5 kg/cm^2 as compared with 2×10^6 kg/cm^2 for steel. The number of fibers per gram will depend on their weight per unit length. Individual fiber weights are in the range of about 1×10^{-6} to 3×10^{-6} g/cm (Corte, 1982). Browning (1970) has calculated that there are from 1 to 10 million fibers in 1 g and that about 1,000 fibers placed side by side in one layer will span 1 in., showing an average fiber width direction span of 25.4 μm in the formed web. In forming the paper web, fibers of various dimensions are arranged in an interlocking network to form a sheet whose structure is determined by the spatial distribution and orientation of the various fiber fractions.

Fibers in paper lie essentially in the plane of the sheet, giving paper a layered or laminar structure but not discrete layers. This results from the thickening and filtration that take place as the water is removed. In the drying stage of the web, removal of the water permits the establishment of hydrogen bonds between the fibers. Hydrogen bonds are now generally accepted as the cause of mechanical coherence of paper (Corte, 1980). Nissan (1983) has pointed out that paper may be treated as a continuum of hydrogen bonds, with the two parameters that determine the modulus of elasticity being the stretch force constants of the hydrogen bonds and the density of such bonds per unit area. He also observes that the hydrogen-bond theory explains the mechanical behavior of paper in terms of independently derived molecular and thermodynamic parameters. If variance is assumed around the mean value of the hydrogen bond, the rupture energy of paper can be related to the number of hydrogen bonds (Nissan, 1983).

Centuries of man's experience with paper have shown that the hydrogen bonds established at the time of manufacture remain intact throughout the life of the paper product, assuming normal storage and use conditions. The strength, integrity, and concentration of the hydrogen bonds in the paper structure help preserve the strength of paper under adverse storage conditions, such as oxidizing atmospheres that degrade the cellulose polymer. On the other hand, if water is reintroduced into the interfiber bond area, it may be absorbed by the hydroxyl groups of cellulose that are associated with the hydrogen bonds of the paper structure. In this way, water can effect reversal of the interfiber bonds and greatly weaken the paper. Although hydrogen bonds are re-established upon removal of the water through drying, they may be displaced, causing cockling of the paper and possible loss of strength. Using this reversible action of water, used paper can be repulped with water and wetting agents and recycled to form paper products using reclaimed pulp. The properties of these pulps are generally lower in strength and brightness than comparable virgin pulp. Contamination by plastics is also a major problem. Reclaimed fibers are not recommended for archival papers.

Many types of nonfibrous raw materials are added to improve the physical, optical, and electrical properties of the resulting paper (Browning, 1970; Clark, 1978). Polymeric binder materials are used to improve the cohesion of the individual fibers and increase the strength and stiffness of the paper. Bonding agents include such materials as starch, modified starches, gelatin, polyvinyl alcohol, methylcellulose, and latex or water emulsion materials such as polystyrene-butadiene, polyacrylates, and polyacrylamides. Inert inorganic materials known as pigments or fillers are added to fill voids between the fibers and to smooth the surface for printing (Hagemeyer, 1984). Fillers also improve the opacity and brightness of the sheet, depending on the particle size, refractive index, and brightness of these materials. Commonly used fillers include clays, talc, calcium carbonate, titanium dioxide, aluminum oxides, and silicates. Pigments are used in varying amounts, depending on the grade of paper, and may comprise 2 to 40 weight percent of the final sheet.

Other additives include sizing agents that are used to reduce the penetration of liquids such as offset printing solutions and fluid printing inks. Rosin, starches, and synthetic resins are examples of sizing materials. The sizing agents may be added as part of the paper raw material to produce internal sizing, or the dry sheet may be passed through a size-press coater that applies a surface size to the sheet.

Rosin is the most widely used sizing agent. The rosin is added to the paper stock with one to three times as much aluminum sulfate, which precipitates the rosin on the fibers as flocculated particles; after addition of the alum, the pH should be 4.5 to 5.5. Sodium aluminate may also be used to precipitate the rosin size, thus attaining slightly higher pH papers. Much attention has been focused on the influence of acid conditions encountered during papermaking on the rate of aging of paper. As a result, the production of paper under neutral to alkaline conditions has gained in importance.

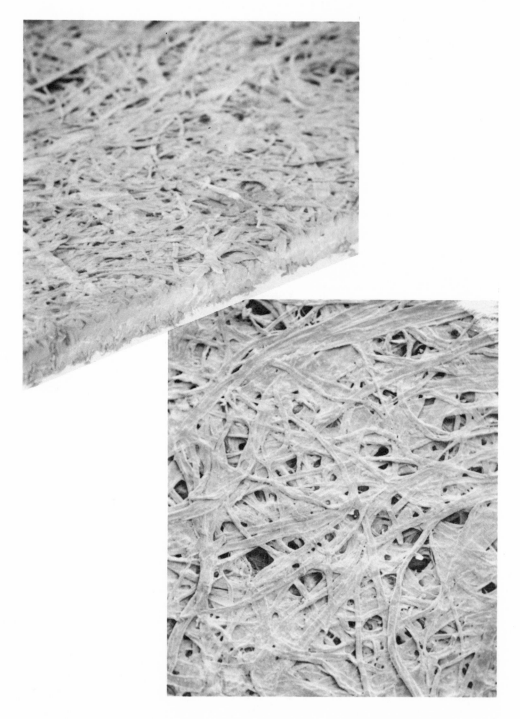

*Photomicrographs of edge and surface of paper showing fiber structure. The
paper is a Nekoosa neutral-pH cotton bond, sub. 20 (76 g/m²), that meets
TAPPI, ASTM, and ANSI requirements for permanence.*

Internal sizing of papers under alkaline conditions (pH 7.0–9.0) is achieved with synthetic sizes such as alkyl ketene dimer and alkenyl succinic anhydride. These sizing agents are combined with calcium carbonate filler to provide a useful pH control by a buffering action during aging of the paper. Recent decreases in the cost of calcium carbonate filler and increases in the cost of virgin pulp have created somewhat more favorable economics for alkaline pH paper, resulting in increased commercial interest and production of this type of paper.

PHYSICAL PROPERTIES

A sheet of paper has been defined as a foil with a fibrous fine structure. It is the structure that determines the physical properties of the paper, and any change in structure affects these properties. The properties required for various types of paper, such as bond, writing, printing, book, envelope, and tablet, are developed by the paper manufacturer through fiber selection and refining, type and amount of additives, manufacturing process parameters, and conversion processes including surface coatings.

The behavior of the paper structure can be shown by typical stress-strain curves for tension, compression, or shear (Setterholm and Gunderson, 1983). In the tensile test, when a piece of paper is subjected to a tensile load it stretches in exact proportion to the applied load, and when the load is released it returns to its original length only if the load does not exceed the elastic limit. If the applied load exceeds the elastic limit of the sample and is released, the piece of paper will contract but not to its original length. The permanent length increase is due to inelastic response of some of the elements or of the fibers that are stretched or straightened in the direction of the load. Under constant load above its elastic limit, paper exhibits viscoelastic creep, and at some maximum value of stress the paper undergoes tensile failure. Typically, elastic response will continue for about one-fourth of the failing stress. In a review of paper behavior, Perkins (1983) presents data to show that in a controlled environment paper will exhibit elastic behavior under low loads of short duration, viscoelastic behavior under low loads of long duration, and inelastic behavior as the level of stress increases. He points out that microfailure could be the result of a variety of processes, including tensile fracture of fibers, failure of fiber-to-fiber bonds, and development of slip planes within the fiber cell walls. The inelastic response of paper is similar to other materials such as reinforced fiber composites. In any event, tensile strength is the force parallel to the plane of the sheet that is required to produce failure in a specimen of specified width and length under specified conditions of loading (Technical Association of the Pulp and Paper Industry [TAPPI] T404 and T494, 1984).

Stretch is the extension or strain resulting from the application of tensile load under specified conditions (TAPPI T404 and T494). The initial slope of the load-elongation curve defines the modulus of elasticity or Young's modulus in the machine or cross-machine direction. Stretch is greatest in the cross-machine direction.

Tearing strength is the average force required to tear a single sheet of paper under standardized conditions (TAPPI T414). Fold endurance is the number of folds a paper can withstand before failure (TAPPI T423 or T511). Brightness is the reflectivity of paper or pulp for light at 457 nm (TAPPI T452). Color is measured by reflectivity according to TAPPI T442 and T524. Opacity relates to the ratio of the diffuse reflectance of the sheet when backed by a black body to that when backed by a white body (TAPPI T425).

It is well understood that temperature has an effect on paper properties and that moisture content has an even larger effect. Pulp and paper are hygroscopic and can absorb water from or lose it to the surrounding atmosphere. As a result, paper properties will change with changes in relative humidity; therefore, the control of relative humidity in the environment for long-term storage of paper records is very important. The influences of water content on the rate of degradation of cellulose have been reported (Graminski et al., 1978) and show that the effect of temperature was less than that of moisture. A molecular layer of water on cellulose occurs at about 5 percent moisture content, and the mechanical strength properties decrease rapidly when the absorbed water exceeds 5 to 7 percent because of the competition between the water molecules and the hydroxyls of cellulose for the hydrogen bonds with other hydroxyls. In addition, Kadoya and Usuda (1984) found that, at 80 percent relative humidity, the fracture mechanism under load changed from bond breaking to fibers sliding out of the network. It is important to carefully control both temperature and relative humidity in pulp and paper test laboratories and to carefully condition test samples before testing (see TAPPI Standard T402-70) to obtain reproducible results.

PERMANENCE FACTORS

Aging studies, by a number of investigators, on papers that have retained their properties over very long periods of time, such as books that have survived for centuries, show clearly the importance of composition in the keeping properties of paper. Hudson's (1976) studies identify the quality of the fiber and the level of acidity as key factors in paper permanence. He shows that there is a good correlation between cold-water-extraction pH and resistance to heat-aging of paper. Through raw material selection and pH control it is possible to make paper that will store for centuries information of importance to civilizations, and this paper permanence can be increased through the use of controlled storage conditions of temperature and humidity consistent with requirements for use. The potential value of cold storage for books and papers that are not in active use was shown by the work of Hudson (1976) and Hudson and Edwards (1966) on books kept in Antarctica from 1912 until 1959 compared with books from the same edition kept in London. Those in the Antarctic were in essentially new condition, while those stored in London showed extensive deterioration.

Nakagawa and Shafizadeh (1984) showed that pure cellulose has a high degree of thermal stability up to 300°C in an inert nitrogen atmosphere. They investigated the rate of change in the molecular weight of cellulose versus time of aging in air at 150°C and 190°C and found that the rate of change decreases with time of aging. This result agrees with earlier findings that the rate of aging in paper under accelerated aging conditions decreases with time. This indicates that the initial rate of aging for paper under controlled storage conditions, as determined by the rate of change in physical properties such as fold endurance, will decline as the sample ages. Browning (1970) reviewed the role of raw material selection and the paper manufacturing process in determining paper durability and the role of both initial composition and storage conditions in determining paper permanence. He pointed out that the permanence of the fiber raw material is governed by the natural resistance of cellulose to deterioration. He also indicated that an initial high paper strength and a low rate of aging as determined by the rate of change of physical properties with time of aging is evidence of a permanent paper. Mendenhall, Kelly, and Williams (1981) showed that the rate of change in fold endurance during aging decreases as the sample ages under stable storage conditions.

Activation energies have been calculated from Arrhenius plots of the rates of change for a number of properties. Thermogravimetric and differential thermal analysis by Ramiah (1970) gave activation energies for oxidation in the range of 36 to 60 kcal/mol for cellulose. Using similar techniques, Cardwell and Luner (1978) produced values ranging from 34.5 to 43.9 kcal/mol for various types of bleached pulp. Browning and Wink (1968) found that Arrhenius plot activation energies of property changes versus time were about the same for a number of paper property tests, including fold, burst, tear, tensile, and a number of specific absorption tests. These investigators suggested that an accelerated aging test of 100°C for 72 hours is equivalent to about 25 years of normal aging at room temperature. The same test conducted with the sample under moist conditions is equivalent to about 300 years of normal aging.

Duswalt (1977) reported an activation energy of 34 kcal/mol ±2 percent and noted that the oxidation rates determined by differential thermal analysis agreed with the ranking of the samples established through changes in fold endurance under accelerated oven aging. Results obtained by Wilson and Parks (1980) found that thermal decomposition of paper correlated well with pH value as determined by a cold-water-extraction test (TAPPI T435).

Extensive tests by Wilson and Parks (1980) on a number of commercial samples tested periodically over a 36–year period of natural aging in comparison with accelerated aging showed an empirical correlation between accelerated aging and natural aging results. The use of accelerated aging tests for the prediction of paper permanence was further investigated and discussed by Wilson and Parks (1983) in a review of the work by the National Bureau of Standards on material for archival records. This work supports the conclusions of other investigators that the development of initial archival quality requires the control of material composition.

Roberson (1976) in his very thorough review of paper permanence and durability pointed out that the changes in paper properties with time are complex and varied. In addition to the pH effect of hydrolysis, which many investigators have proposed as being predominant, he cites oxidation, cross-linking, and the influence of light. Luner (1969) supports the complex nature of aging changes in paper permanence in his review of both physical and chemical interactions. Both Roberson (1976) and Luner (1969) also review the important test methods and test procedures for evaluating paper for permanence characteristics.

Sclawy and Williams (1981) cite a number of reports showing that the pH value of paper is a major factor in developing and controlling paper permanence and that low-pH papers age more rapidly than neutral to alkaline papers. They give examples of ancient papers that have kept for centuries, many of which remain in good condition today. Test results on these papers suggest that the permanence exhibited is due to an alkaline to neutral pH value or an alkaline filler or both. The

Effects of slow-working acid in book paper. Peter Waters, conservation officer at the Library of Congress, demonstrates that a hefty puff makes confetti of deteriorated pages.

benefit of an alkaline reserve is demonstrated by the keeping qualities of paper contained in a book published in 1801. The paper was made by combining straw pulp and groundwood pulp with chalk as a whitening agent. The paper is in good condition in the personal library of J. d'A. Clark (1978). Barrow's (1960) results on old books showed that one book, published in Venice in the 17th century, that had unusual keeping qualities contained an unusually high level of calcium carbonate filler. This suggests that permanent papers can be produced using refined lignin-free wood pulp that has an alkaline filler. Information based on the composition of such well-preserved centuries-old paper samples has influenced favorably the development of standards and specifications for the production of archival-quality papers (see section on Standards and Specifications later in this chapter).

Kelly (1972) has shown that for a given type of paper it is possible to calculate the rate of acid development caused by effects from such conditions as rosin-alum size, contact with acid atmospheres (SO_2, NO_x), and oxidation; also shown is the offsetting effect of an alkaline filler in the sheet. His calculations indicate that the acid generated by the paper-based reactions is neutralized by the excess alkali contained in the sheet. He reviews the possible mechanism for acid generation, including the role of trace metals as oxidation catalysts. As a result, it is proposed that archival paper be made with an alkaline reserve and be free of oxidation catalysts. Both Williams (1979) and Browning (1969) find that the properties required for archival-quality papers are well established. Papers meeting these requirements can be produced using commercially available raw materials, including fiber and filler, and processes that yield neutral alkaline pH in the final sheet. Archival paper records printed on such a paper base will keep for centuries under suitable storage conditions.

PRESERVATION

The problems of paper preservation associated with the low permanence of acid-sized papers have led to extensive investigations of laboratory- and production-scale processes for increasing permanence of existing paper records such as maps, charts, documents, and books. Smith and Wilson (1970) reviewed a number of deacidification procedures, with particular emphasis on a nonaqueous process developed by Smith. This process, involving the treatment of the paper with an organic solvent solution of an alkali or alkaline-earth alkoxide such as magnesium methoxide, has been developed to commercial-scale practice by Smith; the commercial-scale process is described in U.S. Patents 3,676,055 and 3,676,182.

Williams and Kelly have reduced a method of deacidifying paper to practice (U.S. Patent 3,969,549, 1976), and a commercial-scale facility is currently being planned by the Library of Congress. The method involves exposing the paper to the vapors of diethyl zinc followed by in situ hydrolysis of the zinc compound to a mildly basic material. Barrow and Sproull (1959) proposed deacidification of papers by soaking them in a solution of calcium hydroxide followed by a further soak in a solution of either calcium or magnesium carbonate, which leaves calcium carbonate in the paper. Although this method has been widely used by conservators in preserving individual documents, maps, and prints, the wet paper is very fragile and must be handled with extreme care until dry. This process is not practical for large-scale conservation because of the time required and the need for highly trained personnel.

The use of magnesium bicarbonate solutions for paper deacidification has been investigated by Wilson and co-workers (1981) at the National Archives. A recent dry process for deacidification, reduced to practice by Kundiot (as described in U.S. Patent 4,522,843), uses an airborne technique to deposit fine particles of $MgCO_3$ on the paper surface. The effectiveness of the process is not known. This development is in accordance with the work of other investigators reported here regarding the effectiveness of an alkaline reserve, which may be added at any time, in extending the useful life of paper.

The results from a number of laboratories over many years show clearly that deacidification procedures are effective in reducing the rate of aging of paper that was not produced according to archival standards. The type of deacidification procedure selected would depend on the type of paper or population requiring the treatment. It should be noted that deacidification does not add physical strength to papers that are treated. All of the currently available deacidification processes may damage certain documents through heat, pH changes, or solvent effects on some inks. This makes it necessary to examine individual documents to exclude those subject to damage by the process used. Because individual document screening would be very costly, mass deacidification is not recommended for the unbound Archives collection.

A review of the patent literature in the United States and Great Britain by Baer and Hanson (1983) shows a high level of activity in paper deacidification methods and materials, with about 20 patents issued.

In addition to these chemical preservation processes, plastic materials for paper preservation and restoration have been extensively investigated and used. Wilson and Parks (1983) reviewed and described a number of reports of investigations and actual use of lamination and encapsulation. Their review includes discussion of the importance of mechanical protection with better enclosures and alkaline folders. When paper is encapsulated or laminated, even very old and very fragile paper can be safely handled, thereby prolonging its useful life. The development and use of a number of preservation methods and materials was reviewed by Roberson (1981).

STANDARDS AND SPECIFICATIONS

Kelly and Weberg (1981) reviewed paper specifications developed by a number of organizations for permanent or archival papers, including the Library of Congress, the American National Standards Institute (ANSI), the American Society for Testing and Materials (ASTM), the National Bureau of Standards (NBS), the Society of American Archivists, and Barrow Laboratories. The specifications listed for paper permanence include a pH of 7.5 to 10.3, at least a 2 percent calcium carbonate reserve in the paper, and the absence of lignin or groundwood pulp. It is estimated that papers meeting the specification should have a probable life of 500 to 1,000 years under good storage conditions. Work on the development of specifications at the National Bureau of Standards, including literature on the stability of paper, was reviewed by Wilson (1974). A general review of the principles involved in alkaline sizing, which is specified for permanent paper, was presented by Tosh (1981). Sizing agents that operate without alum, such as alkyl ketene dimers or anhydrides, are recommended.

Document encapsulation. Paper that is too fragile for handling can be protected between sheets of transparent plastic film.

Specifications for papers for permanent records have been published by ANSI and ASTM. Test methods applied to paper have been published by the Technical Association of the Pulp and Paper Industry (TAPPI). Each TAPPI standard for permanent paper designates three levels of permanence: Type I, Maximum Permanence, pH 7.5–9.5; Type II, High Permanence, pH 6.5–8.5; and Type III, Medium Permanence, pH 5.5 minimum. In addition, a 2 percent calcium or magnesium carbonate filler level is specified for Type I papers. The ANSI and ASTM standards are designated D3290-76, Bond and Ledger Paper for Permanent Records; D3208-76, Manifold Papers for Permanent Records; D3301-74, File Folders for Storage of Permanent Records; D3458-75, Copies From Office Copying Machines for Permanent Records; and Z39.48-1984, Permanence of Paper for Printed Library Materials.

For the creation of paper records meeting archival quality standards, permanent or archival paper should be used in combination with permanent or archival-quality image-forming materials. Printing inks for printing presses and toners for copying machines are a combination of pigments or colorants with a resin to bind the pigment to the paper surface. A wide selection of both types of raw materials is available to the ink and toner manufacturer, including materials of proved long-term stability such as carbon black and colored stable inorganic pigments and stable types of resin such as polyesters, polyamides, acrylics, and phenolics. Not all inks and toners are designed for long-term stability; therefore, archival requirements should be applied in the selection of these materials for the creation of archival-quality records. In addition, both the printing process and the copying process should be operated under conditions that are optimum for promoting a strong bond between the ink or toner and the paper surface. If these materials are used in well-adjusted processes in combination with archival-quality paper (ASTM D3458-75), the resulting paper record will keep under good storage conditions for many centuries. A useful reference for guidance in selection of archival-quality inks and toners and the xerographic process is the Printing Ink Manual (1979); see also Diamond (1984) and Parks and Wilson (1974).

ADVANTAGES, DISADVANTAGES, AND CONCLUSIONS

Advantages

The advantages of paper as an archival material are these:

1. The permanence of paper generally demonstrated through many centuries of storage in collections throughout the world is an advantage for archival use.

2. The ease of producing copies of paper documents as a result of the worldwide proliferation of copiers and duplicators in offices, libraries, airports, hotels, etc., is an advantage for paper-based records. The copying process has no discernable detrimental effect on the original.

3. Printers or "intelligent" copiers that produce paper copies directly from computer-based records are an advantage in computer-based record systems. These copiers can print a variety of information using computer-generated formats at any location using satellite or telephone-line transmission.

4. The ready availability of neutral pH paper at regular commercial prices for

use in generating archival-quality papers is an advantage today that did not exist a few years ago.

Disadvantages

Paper's disadvantages are as follows:

1. Papers that are acid-sized show degradation during storage, and deacidification may be needed to extend their useful storage life.
2. Many of the early copier and duplicator processes developed fragile or unstable copies that do not have longevity. In some cases the images were composed of dyes that fade with time.
3. Paper-based records are bulky and involve manual operations.

Conclusions

The following conclusions regarding paper are drawn:

1. Experience with use and storage of paper records over many years has demonstrated the centuries-long permanence of paper produced with near-neutral pH made from bleached pulps and with an alkaline reserve.
2. Images formed on permanent paper with inks from printing presses or toners from photocopying machines that use permanent-type materials such as carbon black pigment and inert resin binders (e.g., polyester, silicone, polystyrene, and epoxy) will remain legible for hundreds or thousands of years if protected by suitable storage conditions.
3. The permanence of all papers can be extended through the use of proper environmental conditions such as low temperatures, humidity control, and dark storage.
4. Paper produced with an internal acidity below 5.5 pH and without an alkaline reserve will have a lower degree of permanence, but increased permanence can be achieved with treatment by laboratory or commercial deacidification methods and the introduction of an alkaline reserve at any time during the paper's useful life. Remedial actions include (a) treatment by an alkalization process, such as deposition of magnesium bicarbonate, when the original documents must be preserved and (b) copying onto permanent paper with permanent toners to provide long-term stability when the intrinsic value of a document is not important. This procedure is particularly recommended when only part of a set or random sheets of documents are to be preserved.
5. Papers that are damaged physically or that have become weak from aging effects may be safely protected through encapsulation using acid-free alkaline-reserve permanent papers or inert materials such as Mylar polyester film.
6. Under suitable storage conditions, the rate of aging decreases with time. Paper that is shown by tests to be aging slowly will change to a lower aging rate if stored under proper environmental conditions.
7. Inactive paper records may be safely stored at reduced temperature, including below-freezing cold storage conditions, to extend their useful life.

REFERENCES

Baer, N. S., and K. Hanson. 1983. Survey of Patent Literature Pertaining to Deacidification of Archives and Library Materials. New York: New York University Conservation Center of the Institute of Fine Arts.

Barrow, W. J. 1960. The Manufacture and Testing of Durable Book Papers, R. W. Church, ed. Publication 13. Richmond, Virginia: Virginia State Library.

Barrow, W. J., and R. C. Sproull. 1959. Permanence in book papers. Science, 129:1075–1084.

Browning, B. L. 1969. Analysis of Paper. New York: Marcel Dekker.

Browning, B. L. 1970. The nature of paper. Libr. Q., 40(1):18–38.

Browning, B. L., and W. A. Wink. 1968. Studies on the permanence and durability of paper. Tech. Assoc. Pulp Pap. Ind. J., 51(4):156–163.

Cardwell, R. D., and P. Luner. 1978. Thermogravimetric analysis of pulp; kinetic treatment of dynamic pyrolysis of papermaking pulps. Tech. Assoc. Pulp Pap. Ind. J., 61(8):81–84.

Clark, J. d'A. 1978. Filling and bonding materials. Chapter 31 (pp. 664–678) in Pulp Technology and Treatment for Paper. San Francisco: Miller Freeman Publications.

Corte, H. 1980. Cellulose water interactions. Chapter 1 (pp. 1–89) in Handbook of Paper Science, Vol. 1, H. F. Rance, ed. New York: Elsevier.

Corte, H. 1982. The structure of paper. Chapter 9 (pp. 175–282) in Handbook of Paper Science, Vol. 2, H. F. Rance, ed. New York: Elsevier.

Diamond, A. S. 1984. Toner and Developer Industry Update. Ventura, California: Diamond Research Corp.

Duswalt, A. A. 1977. Thermal analysis study of paper permanence. Chapter 23 in Preservation of Paper and Textiles of Historic and Artistic Value, J. C. Williams, ed. Am. Chem. Soc. Adv. Chem. Ser. 164.

Emerton, H. W. 1980a. The fibrous raw materials of paper. Chapter 2 (pp. 91–138) in Handbook of Paper Science, Vol. 1, H. F. Rance, ed. New York: Elsevier.

Emerton, H. W. 1980b. The preparation of pulp fibers for paper making. Chapter 3 (p. 139) in Handbook of Paper Science, Vol. 1, H. F. Rance, ed. New York: Elsevier.

Graminski, E. L., E. J. Parks, and E. E. Toth. 1978. The effects of temperature and moisture on the accelerated aging of paper. Pp. 341–355 in Durability of Macromolecular Materials, R. K. Eby, ed. Am. Chem. Soc. Symp. Ser. 95.

Hagemeyer, R. W., ed. 1984. Pigments for Paper. Tech. Assoc. Pulp Pap. Ind. Papers.

Hudson, F. L. 1976. The permanence of paper. Pp. 714–723 in The Fundamental Properties of Paper Related to Its Uses, F. Bolam, ed. Trans. Fundam. Res. Symp., Cambridge, 1973. London: Technical Section of the British Paper and Board Makers Association.

Hudson, F. L., and C. J. Edwards. 1966. Some direct observations on the aging of paper. Pap. Technol., 7(1):27–31.

Hunter, D. 1978. Papermaking—The History and Technique of an Ancient Craft. New York: Dover Publications, p. 5.

Kadoya, T., and M. Usuda. 1984. The penetration of non-aqueous liquids. Chapter 19 (pp. 123–141) in Handbook of Physical and Mechanical Testing of Paper and Paperboard, Vol. 2, R. E. Mark, ed. New York: Marcel Dekker.

Kelly, G. B. 1972. Practical aspects of deacidification. Bull. Am. Inst. Conserv., 13(1):16–28.

Kelly, G. B., and N. Weberg. 1981. Specifications and test for alkaline papers. Pp. 71–76 in Technical Association of the Pulp and Paper Industry Papermakers Conference Proceedings (April). Atlanta: TAPPI Press.

Luner, P. 1969. Paper permanence. Tech. Assoc. Pulp Pap. Ind. J., 52(5):769–805.

McGovern, J. N. 1978. Pulp Paper, 52(9):112.

Mark, R. E., ed. 1983. Mechanical properties of fibers. Chapter 10 (pp. 409–495) in Handbook of Physical and Mechanical Testing of Paper and Paperboard, Vol. 1. New York: Marcel Dekker.

Mendenhall, G. K., G. B. Kelly, and J. C. Williams. 1981. The application of several empirical equations to describe the change of properties of paper on accelerated aging. Pp. 177–188 in Preservation of Paper and Textiles of Historic and Artistic Interest, Volume II, J. C. Williams, ed. Am. Chem. Soc. Adv. Chem. Ser. 193.

Nakagawa, S., and F. Shafizadeh. 1984. Thermal properties. Chapter 23 (pp. 241–279) in Handbook of Physical and Mechanical Testing of Paper and Paperboard, Vol. 2., R. E. Mark, ed. New York: Marcel Dekker.

Nissan, A. H. 1983. Retrospect and prospect of testing. Chapter 1 (pp. 1–19) in Handbook of Physical and Mechanical Testing of Paper and Paperboard, Vol. 1, R. E. Mark, ed. New York: Marcel Dekker.

Parks, E. J., and W. K. Wilson. 1974. Evaluation of Archival Stability of Copies From Representative Office Copying Machines. National Bureau of Standards Report No. 74-498(R), April 30.

Perkins, R. W. 1983. Elastic, viscoelastic, and inelastic behavior. Chapter 2 (pp. 23–75) in Handbook of Physical and Mechanical Testing of Paper and Paperboard, Vol. 1., R. E. Mark, ed. New York: Marcel Dekker.

Printing Ink Manual, 3rd edition. 1979. London: Northwood Books.

Ramiah, M. V. 1970. Thermogravimetric and differential thermal analysis of cellulose, hemicellulose and lignin. J. Appl. Polym. Sci. 14(5):1323–1337.

Roberson, D. D. 1976. The evaluation of paper permanence and durability. Tech. Assoc. Pulp Pap. Ind. J., 59(12):63–69.

Roberson, D. D. 1981. Permanence/durability and preservation at the Barrow Laboratory. Pp. 45–55 in Preservation of Paper and Textiles of Historic and Artistic Interest, Volume II, J. C. Williams, ed. Am. Chem. Soc. Adv. Chem. Ser. 193.

Rollins, M. L. 1965. The cotton fiber. Pp. 44–79 in The American Cotton Handbook, 3rd edition, Vol. 2, D. S. Hamby, ed. New York: Interscience.

Sclawy, A. C., and J. C. Williams. 1981. Alkalinity, the key to paper "permanence." Tech. Assoc. Pulp Pap. Ind. J., 64(5):49–50.

Setterholm, V. C., and Gunderson, D. C. 1983. Observations on load deformation and testing. Chapter 4 (pp. 115–143) in Handbook of Physical and Mechanical Testing of Paper and Paperboard, Vol. 1., R. E. Mark, ed. New York: Marcel Dekker.

Sjostrom, E. 1981. Wood Chemistry Fundamentals and Applications. New York: Academic Press.

Smith, R. D., and W. K. Wilson. 1970. New approaches to preservation of library materials. Libr. Q., 40(1):139–175.

Technical Association of the Pulp and Paper Industry. 1984. Combined Test Methods Manual. Atlanta, Georgia: TAPPI Press.

Tosh, C. 1981. Durability of paper. Paper (London), 195(9):26–30.

Williams, J. C. 1979. Paper permanence: A step in addition to alkalization. Restorator, (3):81–90.

Wilson, W. K. 1974. Development of Specifications for Archival Record Material. National Technical Information Service Report COM-75-10131.

Wilson, W. K., and E. J. Parks. 1980. Comparison of accelerated aging of book papers in 1937 with 36 years of natural aging. Restaurator, (4):1–55. Copenhagen: Munksgaard.

Wilson, W. K., and E. J. Parks. 1983. Historical survey of research at the National Bureau of Standards on materials for archival records. Restaurator, (5):191–241. Copenhagen: Munksgaard.

Wilson, W. K., R. A. Golding, R. H. McCloren, and J. L. Gear. 1981. The effect of magnesium bicarbonate solutions on various papers. Pp. 87–107 in Preservation of Paper and Textiles of Historic and Artistic Interest, Volume II, J. C. Williams, ed. Am. Chem. Soc. Adv. Chem. Ser. 193.

Film storage facility in Granite Mountain records vault of the Genealogical Society of Utah. Control of storage conditions is vital to ensure permanence of stored records.

5

Photographic Film

Photography has been in existence for over 150 years, and the basic principle has remained essentially unchanged for black-and-white images. A photographic image is formed by the action of light on silver halide salts contained in a binder applied to a plastic substrate. This allows the exposed silver halide to be converted to metallic silver by the subsequent action of a photographic developer, yielding a reversal or negative image. The unexposed silver salts are removed in the photographic processing. Other technologies are applicable for producing the photographic image, but silver halide salts are by far the most used for recording black and white images.

STRUCTURE

The film's light-sensitive layer is called the photographic emulsion; it consists primarily of silver halide in a colloidal medium. The halide compound may be silver chloride or bromide or mixtures of the chloride, bromide, or iodide. The medium is usually gelatin, although the very early photographic materials employed albumen or collodion. The photographic emulsion also contains optical sensitizers, antifoggants, and other chemicals with specific purposes, such as hardening. Photographic emulsions vary in thickness but are generally from one to several ten-thousandths of an inch. Color emulsions are thicker than black-and-white emulsions.

To provide the required mechanical integrity and to permit easy handling, photographic emulsions are coated onto a transparent base or support. Originally, glass was used as the major support when negatives were made on photographic plates. Photographic film came about because of the availability of flexible plastic supports. Film supports range in thickness from 0.0025 to 0.008 in. (65.5 to 215 μm), depending on the chemical nature of the support and the end-use requirements. The first plastic film support was cellulose nitrate, which was in commercial use from about 1890 to 1950. It was one of the first plastics in existence, and its

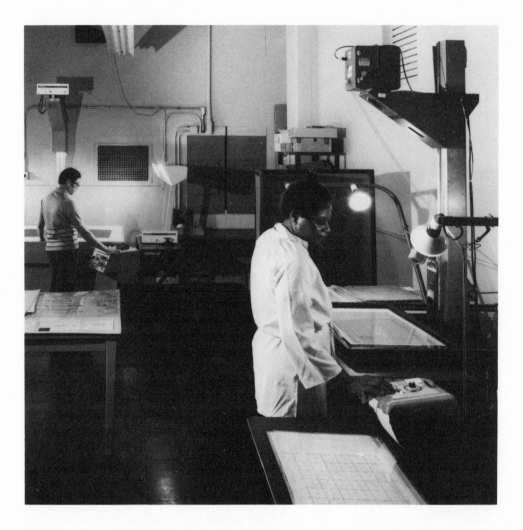

Microfilming laboratory at the National Archives. Careful inspection and quality control procedures assure integrity of records.

application marked the beginning of the widespread use of photography. It proved to be a very acceptable film base but suffered from high flammability and poor chemical stability. These drawbacks were recognized early, and there was considerable research after World War I to obtain a more stable material. In the 1920s and 1930s, other derivatives of cellulose (i.e., cellulose acetate butyrate and cellulose acetate propionate) found commercial application for certain product types, primarily amateur and X-ray films. After World War II, cellulose triacetate was developed, and the material replaced cellulose nitrate completely. The latter ceased to be manufactured in the United States for photographic film use early in the 1950s.

While cellulose triacetate and other cellulose esters overcame the concerns about chemical stability, they were not the ideal film supports for all end uses. Many products require extremely good dimensional stability, primarily in the graphic arts and aerial photography fields. To meet this need, polyethylene terephthalate was introduced around 1956. It is commonly known as polyester film base, and it has since found very wide acceptance. Today most photographic films are coated onto either cellulose triacetate or polyester base. It is these two materials that must be considered for use by the National Archives.

The cellulose ester supports are manufactured by the solvent-casting process. The cellulose ester, together with plasticizer, is dissolved in solvents. This viscous fluid is then spread evenly on a huge, highly finished casting drum or endless moving belt. The solvent is evaporated, and when the cast support is firm enough it is stripped from the drum or belt and is cured further in heated chambers.

Polyester support is manufactured by a completely different process, since it is insoluble in all common solvents. It is melted and extruded through a precision-finished die to form a continuous sheet. To obtain the necessary physical properties, it is subsequently stretched in both the machine and cross directions at elevated temperature. Finally, it is subjected to a temperature considerably above the stretching temperature while mechanically restrained to prevent shrinkage. This annealing causes the polymer chains to crystallize and imparts thermal stability to the support.

Film supports are coated with various layers during manufacture to provide good adhesion with the photographic layer, to give static protection, or to eliminate light reflecting from the support surfaces back into the light-sensitive emulsion.

The photographic emulsion is coated onto the plastic support from aqueous solutions. The liquid emulsion flows onto the support from a coating hopper, and the water is then evaporated in curing sections. Depending on the product, the total emulsion may consist of one or many individual layers. Color emulsions have one or two layers for each color, along with filter layers. There can be as many as a dozen layers. Black-and-white archival emulsions have a total of one or two layers.

APPLICATIONS

Photography has been used to copy documents since the 1930s, when banks made photographic copies of checks. This approach has been widely expanded, and reduction printing is used for many types of documents, giving rise to the microfilm industry. There are multiple purposes for microfilming, as described by

Gille (1953), "to preserve documents from eventual and non-voluntary destruction (fire, war, etc.); to permit the communication (in the form of copies) of particularly valuable or fragile documents; to facilitate the consultation of documents of special formats or volume; to enable the archives to satisfy requests for consultations of records outside the archives by sending copies."

An additional purpose was described by Leisenger (1976) of NARS: ". . . the National Archives outlined a program for the rehabilitation of the damaged records. It was decided to preserve records by microfilming them whenever feasible and to reserve the more costly laminating and rebinding processes for those records that are too fragile for microfilming, that have special values requiring their preservation in their original form."

Crespo Nogueira (1982) further identified the different purposes of filming an archive. She distinguished "reference microfilming, security microfilming, acquisition or complementary microfilming, space saving or substitution microfilming, and publications microfilming. In most cases there will be no clear-cut separation between these various uses, which will frequently overlap. Whatever the primary reason for a microfilming operation may be, it will in most cases also serve some if not all of the other possible purposes. Undoubtedly, considering the various uses of microfilming, documentary preservation is the most frequent to appear as a subsidiary purpose." In the context of the current NARA concerns, substitution microfilming would be considered a preservation measure because of the poor physical condition of the original documents and the high possibility of their eventual loss. Crespo Nogueira also noted that "the usefulness of microfilming as a method of preservation has been clear to archivists from the very beginning of archival microfilming, even if economic reasons have delayed its effective application. Sometimes, however, microfilm may be used as a substitute for more costly restoration which will then be restricted to the more fragile or most valuable documents."

Crespo Nogueira conducted an extensive survey of the microfilming practices of the national archives in various countries. Among the 48 replies received, only four indicate that a microfilming policy does not exist. In 39 countries, the primary purpose of microfilming was archival preservation.

PERMANENCE

The permanence of photographic film is dependent on the stability of the film support, the image, the adhesive layer, and the adhesion between them. Each is here discussed in turn.

The stability of the film support obviously varies with the support type. The chemical instability of cellulose nitrate has already been referred to, but this material has not been used as a film support in this country for over 30 years. The only concern today is that cellulose nitrate-base film is a fire hazard and should be stored accordingly. It should not be stored in the same storage areas as film having long-time value. Any degradation of the cellulose nitrate base can release nitrogen dioxide, which has a devastating effect on the silver image (Carroll and Calhoun, 1955).

The chemical stability of cellulose triacetate and polyester supports and films has been studied using high-temperature aging results to predict lower tempera-

TABLE 5-1 Predicted Life of Photographic Film at 70°F (21°C),
50 percent Relative Humidity

Base	Life	Reference
Cellulose triacetate	> 300 years	Adelstein and McCrea (1981)
Polyester	> 2,000 years	Adelstein and McCrea (1981)
Polyester	> 1,000 years	Brown et al. (1984)

ture behavior (Adelstein and McCrea, 1981; Brown et al., 1984). It was found that initial stability was followed by the change in viscosity and in mechanical properties. These data were treated using the classical kinetic approach of Arrhenius for first-order chemical reactions, although it was well recognized that the change in mechanical properties involved many interactions that are much more complex. Nevertheless, the Arrhenius treatment allowed extrapolation of laboratory data to room-temperature storage conditions.

Predictions by Adelstein and McCrea (1981) and Brown et al. (1984) of the useful life of photographic film at recommended storage conditions are given in Table 5-1. These predictions are believed to be conservative because the property loss on which they were based did not represent a degree of degradation that rendered the film completely unusable. Arrhenius projections for emulsion decomposition were found under the same high-temperature incubations to be in excess of 2,000 years.

The stability of the photographic image depends on the emulsion type, and these types may include black-and-white, color, diazo, vesicular, and thermally processed silver. Of primary concern to the archivist are black-and-white images formed by silver halide. However, it is very difficult to obtain good Arrhenius projections for this type of image because the photographic changes occurring are so small. For example, a black-and-white microfilm image was still very usable after 1,000 days of incubation at 70°F (21°C) and 50 percent RH (Adelstein and McCrea, 1977). The resulting density changes were very small, and Arrhenius relationships cannot be obtained from such data. The only Arrhenius plots for a change in black-and-white image density were reported for X-ray films by Kopperl et al. (1982). These data indicated that a density change of 0.05 in the low-density or clear area for normally processed X-ray film would not be expected to occur within at least several centuries at recommended storage conditions.

The stability of color images is quite different, and the utility of the Arrhenius plot to predict the fading of color images has been known for some time (Adelstein et al., 1970; Bard et al., 1980; Seoka et al., 1982). One obvious solution for preservation of color images is low-temperature (35°F, or about 2°C) storage, although this creates problems in the handling of the materials and is also expensive. It is not a practical or necessary consideration for the National Archives problem.

Diazo films are used in the micrographic industry as a print material; these are dye images. Arrhenius projections have been useful in predicting the life of these products, and one study (Adelstein and McCrea, 1977) predicted useful life ranging from 10 to over 100 years under recommended storage conditions, depending on the product type. These products are considered medium-term or long-term films but are not suitable for archival storage.

Vesicular films have images composed of small bubbles of nitrogen formed by

Mircofilm cabinets in National Archives microfilm reading room.
Self-service microfilm storage cabinets allow researchers
ready access to archival materials.

the decomposition of a diazonium salt. These films are used in the microfilm industry as a print material. Although good experience has been obtained with their image stability (Ram and Potter, 1970; Potter and Ram, 1975), it is not possible to use Arrhenius relationships to predict the eventual life of these materials. Long-time behavior at room-temperature conditions cannot be predicted from short-term tests at elevated temperatures for this material. At the higher temperatures, the binder containing the nitrogen bubbles softens and the bubbles collapse, thereby destroying the image. Because of lack of knowledge of the eventual life of these films, they are not considered an archival medium.

Thermally processed silver films are also used in the micrographics industry. A silver salt on a long-chain fatty acid is the image-forming material, and after exposure to light a silver image developed by heat is obtained from the decomposed silver salt. Studies on the permanence of these images (Kurttila, 1977) predict a useful life varying up to 100 years. These films differ from conventional black-and-white silver images in that the latter have unwanted by-products removed during processing. Thermally developed silver films are not considered suitable as an archival medium.

In addition to the stability of the film support and the photographic image layer, it is essential that there be good adhesion between the two and that this adhesion not deteriorate upon aging. Two adhesion tests are described in the American National Standards Institute specifications for archival film (ANSI/ASC PH1.28 and PH1.41). A tape stripping test measures the existing adhesion, and a humidity cycling test measures the stability of the adhesion after repeated cycles. This condition is particularly severe on adhesion properties. All photographic films that are used for archival storage should pass these specifications.

STORAGE STANDARDS

The storage life of photographic film is dependent on the interactions among three separate factors: (a) the film type, (b) the photographic processing, and (c) the storage conditions. Both American national and international standards are now available that cover all three of these factors. Each is discussed here in turn.

Inasmuch as this report deals primarily with the preservation of historical records and not with distribution of information on film, only black-and-white silver halide film will be considered. It is the only film type that is considered to be archival, i.e., suitable for the storage of information having permanent value. Other types of film (diazo, vesicular, etc.) are used only for producing inexpensive copies of the negative film. All films in use today must pass the requirements for safety film, to distinguish them from the older flammable and unstable nitrate-base film. The following approved American National Standards are applicable to archival films: ANSI PH1.25-1984, for Photography (Film)—Safety Photographic Film; ANSI/ASC PH1.28-1984, for Photography (Film)—Archival Records, Silver-Gelatin Type, on Cellulose Ester Base; and ANSI/ASC PH1.41-1984, for Photography (Film)—Archival Records, Silver-Gelatin Type, on Polyester Base.

Processing of exposed film can be successfully done in many different manufacturers' brands of processors. There are procedures that must be followed to ensure that archival film is obtained. The specifications for maximum levels of residual silver halide salts and processing chemicals are given in ANSI/ASC specifications PH1.28 and PH1.41. The most critical processing chemical is the thiosulfate ion, which can cause image staining if at too high a level. It can be determined by the methylene blue method as described in ANSI/ASC PH4.8-1985, Residual Thiosulfate and Other Chemicals in Film, Plates, and Papers—Determination and Measurement.

Archival storage conditions are absolutely essential for the films. The full potential of storage life will only be realized if the film is stored under optimum conditions. These conditions are under the control of the film user, and the user's role in film preservation must be understood by all who are concerned with the film life. The only real disadvantage to photographic film is the necessity for a controlled environment for archival storage. The average temperature should not be over 68°F (20°C), and the relative humidity should ideally be between 30 and 40 percent. Silver halide film should be stored by itself in a secure controlled environment. If it is stored with other material, particular care should be taken to remove all materials that give off contaminants. The air should be filtered for particle contaminants, and harmful gases should be removed or neutralized. Examples of harmful gases are nitrogen dioxide, peroxides, ammonia, sulfur dioxide, and some paint fumes that may cause deterioration of the base and a chemical degradation of the photographic image. Adverse experience has been reported when photographic images are exposed to nitrogen dioxide (Carroll and Calhoun, 1955). This is of practical importance in that nitrate-base films should never be stored in the same area as films having archival value. Another adverse reaction to air pollutants is the appearance of microspots or microblemishes when film is exposed to ozone, peroxides, or oxidizing gases (McCamy, 1964; Henn et al., 1965). Although changes in photographic processing have resulted in film becoming much more resistant to this type of defect, care must be exercised so that photographic images are not stored in an oxidizing atmosphere (Weyde, 1972).

These concerns about the environment also apply to other storage media and are discussed in greater detail in Chapter 3. Even though proper storage is a necessity for archival permanance of film, it is achievable with presently known technology and readily available machinery. In addition to the storage environment, there are specifications for enclosure materials (cartons, cans, jackets, etc.) that are stored with the film. The following standards must be followed: ANSI PH1.43-1985, for Photography (Film)—Storage of Processed Safety Film; and ANSI PH1.53-1984, for Photography (Processing)—Processed Films, Plates, and Papers—Filing Enclosures and Containers for Storage.

In addition to these standards on film material, photographic processing, and storage, there are also the following standards on the formats used to make reduced photographic copies on microfilm and the required inspection and quality control: ANSI/NMA MS23-1983, Practice for Operational Procedures—Inspection and Quality Control of First-Generation, Silver-Gelatin Microfilm of Documents; and ANSI/NMA MS14-1978, Specifications for 16 and 35-mm Microfilms in Roll Form.

Great care has been taken to ensure that manufacturers, users, industry associations, and standards bureaus have all had appropriate input into these standards. Consequently, they represent the best information available and should be used as the foundation for any National Archives filming program.

MICROFILM USES

The versatility of microfilms makes them easily adaptable to various archival needs. The choice of format may be dictated by many factors:

1. Size and volume of original records
2. Physical condition, binding, and storage method of original records
3. Frequency of use of records
4. Needs or demand for distribution of information
5. Economics and money available
6. Equipment and manpower available

The most economical format is roll film, particularly 16-mm when using a high-reduction filming. The roll is less expensive to purchase, and filming takes only half the time of a step-and-repeat camera, which uses a fiche (i.e., sheet film) format. Large files of records with infrequent research use are best suited for the roll format (Powell, 1985). The microfiche format is usually best when the use level is high and large numbers of copies are to be distributed. The other advantage of fiche is the ability to have a complete publication (project or file) on one fiche. Fiche readers are less expensive than roll-film readers.

With the versatility and flexibility of microfilm, uses are limited only by one's ingenuity and/or needs. Filmed images can be used as originally filmed or changed. For example, if roll film is used, it can later be put in fiche format by specialized equipment that cuts and jackets the film to form microfiche. If the information on film warrants the expense of optical disk, then the film can be scanned and the image also put on optical disk.

Film can be indexed for quick retrieval. The index can be completed manually and then filmed for preservation, security, and distribution, or it can be done on computer and distributed on computer-output microfilm. It is recommended that documents to be filmed have an image or frame number and a blip that can be read (counted) electronically to facilitate eventual indexing.

Microfilm has the advantage of data compaction. Depending on the format used, the storage space saved over that needed for the original documents ranges from 92 to 95 percent. This is illustrated in Table 5-2, which shows the average number of pages in a microfilm roll and microfiche.

TRENDS

The present trend in the micrographics industry is toward 16-mm film and away from 35-mm film.

There is also a definite trend toward computer-aided retrieval. An index is made with the help of computer technology, and the desired file or frame location is reported to the researcher. This can be an "on-line" system or a fiche produced from computer-output microfilm.

A new technique for enhancement of images that are not easily readable is reportedly being prepared for commercial availability.

There continues to be a despecialization in the use of film. At one time only specialists filmed and processed microfilm. Now micrographics is being used in an

TABLE 5-2 Format and Compaction of Microfilm

Film Size	Reduction	Number of Pages per Roll[a]	Format[b]
Roll film[c]			
16 mm × 125 ft	24 X	6000[d]	Cine mode, double page
	24 X	3000[d]	Comic mode, single page
	42 X	6000[d]	Cine mode, double page
	42 X	3000[d]	Comic mode, single page
35 mm × 125 ft	15 X	2000[e]	Comic mode
	19 X	2200	
Fiche			
4 × 6 in.	24 X	98	
	48 X	420	

[a]Pages filmed simplex (one document at a time) comic format, 2 (8½ × 11 in.) pages per exposure.

[b]Cine: Documents filmed with pages or frames arranged as in a motion picture film with the writing perpendicular to the edges of the film; Comic: Documents filmed with images running with top of the document at top of film and writing from left to right.

[c]Thickness of polyester-base film is 0.004 in.

[d]The horizontal space on 16-mm film is the same for both 24 X and 42 X reduction, as the cameras have a fixed film advance. The advantage of higher reduction filming is that single large documents or two smaller documents can be filmed in a single exposure.

[e]Additional frames can be realized on 35-mm film by filming in cine mode.

office environment along with word processors, filing cabinets, etc., by regular office personnel.

ADVANTAGES AND DISADVANTAGES

Advantages

Film has a number of advantages as an archival material:

1. Except for physical details such as color notations, watermarks, etc., microfilm preserves the intellectual content (the written word) in the original form because it is an exact picture of the original.

2. The life of microfilms is predictable, provided that present standards are met.

3. There is significant compaction of data.

4. Copies of microfilm are easily and economically made on silver, diazo, or vesicular film.

5. Retrieval of film is easily accomplished. If the system is correctly designed, retrieval of any roll or fiche should be less than 5 minutes.

6. The micrographics industry is a mature industry, with accepted standards already in existence.

7. Micrographics is not a high-technology industry requiring rapidly outdated hardware that is software- or computer-dependent; quite simple optical devices can be used to read the film.

Disadvantages

The disadvantages of film are as follows:

1. Microfilm can be damaged or destroyed by storage at high humidity.
2. Microfilm can be damaged by gaseous pollutants in the storage environment.
3. If the original documents are brittle or damaged, it may be difficult or impossible to film them.
4. Copying onto microfilm requires verification as to photographic quality and content.
5. Microfilm can be damaged by careless handling.

REFERENCES

Adelstein, P. Z., and J. L. McCrea. 1977. Dark image stability of diazo film. J. Appl. Photogr. Eng., 3(3, Summer):173–177.

Adelstein, P. Z., and J. L. McCrea. 1981. Stability of processed polyester base photographic films. J. Appl. Photogr. Eng., 7(6, December):160–166.

Adelstein, P. Z., C. L. Graham, and L. E. West. 1970. Preservation of motion-picture color films having permanent value. J. Soc. Motion Pict. Telev. Eng., 79(November):1011–1018.

Bard, C. C., G. W. Larson, H. Hammond, and C. Packard. 1980. Predicting long-term dark storage dye stability characteristics of color photographic products from short-term tests. J. Appl. Photogr. Eng., 6(April):42–45.

Brown, D. W., R. E. Lowry, and L. E. Smith. 1984. Prediction of the Long Term Stability of Polyester-Based Recording Media. NBSIR 84-2988 (December). Gaithersburg, Maryland: U.S. Department of Commerce, National Bureau of Standards.

Carroll, J. F., and J. M. Calhoun. 1955. Effect of nitrogen oxide gases on processed acetate film. J. Soc. Motion Pict. Telev. Eng., 64(September):501–507.

Crespo Nogueira, C. 1982. The use of microfilm as a means of archival preservation. Pp. 3–8 in Proc. 21st Int. Conf. Round Table on Archives, Kuala Lumpur (November–December 1982).

Gille, B. 1953. Esquisse d'un plan de normalisation pour le microfilmage des archives. Archivim III:87–103. Paris: International Council on Archives.

Henn, R. W., D. G. Wiest, and B. D. Mack. 1965. Microscopic spots in processed microfilm: The effect of iodide. Photogr. Sci. Eng., 9(3, May–June):167–173.

Kopperl, D. F., G. W. Larson, B. A. Hutchins, and C. C. Bard. 1982. A method to predict the effect of residual thiosulfate content on the long-term image-stability characteristics of radiographic films. J. Appl. Photogr. Eng., 8(2, April):83–89.

Kurttila, K. R. 1977. Dry silver film stability. J. Microgr., 10(3, January):113–117.

Leisenger, A. H. 1976. Report of the microfilming committee of the International Council on Archives. Archivim XVI:140. Paris: International Council on Archives.

McCamy, C. S. 1964. Inspection of Processed Photographic Record Films for Aging Blemishes. National Bureau of Standards Handbook 96 (January 24).

Potter, E. W., and A. T. Ram. 1975. Stability of vesicular microfilm images III. Paper presented at Annual Society of Photographic Science and Engineers Conference, Denver, Colorado, May 1975.

Powell, T. F. 1985. The miracle of microfilm: The foundation of the largest genealogical record collection in the world. Microform Rev., 14(3, Summer):148–156.

Ram, A. T., and E. W. Potter. 1970. Stability of vesicular microfilm images II. Photogr. Sci. Eng., 14(4, July):283–288.

Seoka, Y., S. Kubodera, T. Aono, and M. Hirano. 1982. Some problems in the evaluation of color image stability. J. Appl. Photogr. Eng., 8(2, April):79–82.

Weyde, E. 1972. A simple test to identify gases which destroy silver images. Photogr. Sci. Eng., 16(4, July–August):283–286.

Magnetic tape storage area. Vast amounts of data can be preserved compactly by using magnetic recording media.

6

Magnetic Recording Media

Magnetic recording media, in the form of tapes and disks, are by far the most common machine-readable storage media in use today. It is estimated that the U.S. government alone at present uses over 15 million reels of half-inch computer tape. In 1985, more than 40 million video cassette recorders were manufactured worldwide, as well as some 200 million half-inch video cassettes. Today, over 1 billion computer flexible disks are produced annually.

ARCHIVAL CRITERIA

The only other mass memory medium used on such a massive scale is, of course, photographic film, and it is therefore natural to make comparisons between the two. Photographic film is the result of more than 150 years of technical development, and today it is a certifiable archival storage medium even though many early films (e.g., cellulose nitrate-base film) were patently unstable. Magnetic tape has evolved over the past 50 years into a reliable, stable storage medium despite the problems of its early forebears (e.g., vinyl acetate-base film); however, it has not yet been awarded archival status.

This discussion reviews the principal reports on magnetic tape stability published within the past 10 years. Generally, the conclusion derived is that a good-quality tape, stored in the proper environment (i.e., 65°F or 18°C, 40 percent RH) and accorded careful mechanical handling, is likely to remain usable for more than 20 years. The period of 10 to 20 years is of particular significance for all machine-readable records because it is also the useful life expectancy of the hardware itself. Today's electronic equipment (e.g., earth satellites, computers, television receivers, and tape recorders) are not expected to remain in service for more than 10 to 20 years. Two important conclusions stem from this fact: first, the recording media may well outlast the hardware; and second, it will become necessary to recopy the tape record every 10 to 20 years on an ever-changing, probably incompatible, new machine with a new format. This operation, termed file conversion, carries with it, of course, the potential of ever-increasing data compaction. Concomitantly, the financial burden of file-converting the entire archival collection perhaps five or six times per century is likely to be out of the question except for relatively small collections that have great historical importance, sustain heavy use, or require rapid access.

In machine-readable records, any realistic discussion of the archival properties cannot be separated from questions on the longevity of their associated hardware or machines. The long-term stability of the recording medium is necessary, but it is not a sufficient criterion for its use. With human-readable records, on the other hand, the long-term stability of the medium is necessary and sufficient.

For these reasons, it must be concluded that magnetic recording media and other machine-readable recording media (e.g., magneto-optic and optical disks) cannot be recommended for long-term (say, over 25 years) archival applications. Similar conclusions have been put forth by the NARS committee (National Archives and Records Service, 1984) and Mallinson (1985a).

DEFINITIONS

In machine-readable records—i.e., magnetic computer, audio, and video tape, magnetic and optical disks, and phonograph records—it is understood that the recorded information can be usefully recovered only by converting it to a human-readable form such as paper text, a photograph, or a video terminal display. In analog recordings, this conversion requires appropriate hardware, and in digital recordings it requires hardware, software, and documentation. On the other hand, the information in human-readable records is comprehensible simply by visual inspection of the record or a magnified image of the record. Only simple optical hardware, such as microscopes and projectors whose design principles need never change, are required to read the record completely. ⸻

THE ELECTRONIC INFORMATION AGE

Modern civilizations are now entering the so-called Information Age, wherein the vast majority of their information and records are stored, manipulated, and disseminated by electronic means such as computer networks, earth satellite relays, and television broadcasting. It seems that the archival community tends to forget that the principal motive for these technologies is their speed of access and that this speed is only achieved at an extremely high cost. The machines themselves (e.g., computers, satellites, and television receivers) are rarely expected to have a useful life in excess of 10 years. The machine-readable records are operated at ever-increasing information storage densities, not only to store more information but also to decrease access times. This is a trend that is surely inimical to long-term archival preservation.

Since 1956 no less than eight differing, incompatible videotape formats of increasing storage density have emerged. In fact, coincidentally, since 1952 eight differing computer tape formats have been used. The sixteen tape formats are listed in Table 6-1. Each format typically mandates a different machine with its unique set of demodulators, decoders, and reformatters. This proliferation of incompatible systems is the root cause of the archivist's dilemma in adopting

TABLE 6-1 Video and Computer Tape Formats

Product	Current Status
Video tape formats since 1956	
2-inch quadruplex	Obsolete
2-inch quadruplex, double density	Obsolete
1-inch helical, type A	Obsolete
1-inch helical, type B	
1-inch helical, type C	
3/4-inch helical, U-Matic	
1/2-inch helical, Beta-max	
1/2-inch helical, VHS	
1/2-inch computer tape formats since 1952	
7-track NRZI, 100 BPI	Obsolete
7-track NRZI, 200 BPI	Obsolete
7-track NRZI, 556 BPI	Obsolete
7-track NRZI, 800 BPI	Obsolete
9-track NRZI, 800 BPI	
9-track PE, 1,600 BPI	
9-track GCR, 6,250 BPI	
18-track NRZI, 19,000 BPI	

KEY: NRZI = Non-Return to Zero Inhibit; PE = Phase Encoding; GCR = Group Code Recording.

machine-readable records. The speed of access and the electronic data processing abilities are indeed attractive, but it must be recognized that the records and their associated hardware will become obsolete within a couple of decades.

Since the information and communication industries are most definitely not driven by long-term archival considerations, it seems futile to expect technology to resolve this problem. Advances in technology continue to cause the machine-readable problem, and obviously these advances will not solve the problem.

ARCHIVAL PROPERTIES OF MAGNETIC RECORDING MEDIA

Magnetic recording media are made up of three components: the substrate, the magnetic particles or grains, and the binder system.

In rigid computer disks (hard disks), the substrate is an aluminum alloy. The magnetic particles are gamma-Fe_2O_3, and the binder system is usually one of the epoxy family. Because of the relatively high cost of storing data on rigid disks (10^{-3} cents per bit versus 10^{-6} cents per bit on tape), rigid disks are rarely considered for archival applications and, therefore, will not be discussed further. An additional factor against its archival use is the fact that the majority of today's large hard disk files (Winchester drives) cannot be physically separated from the head-disk assembly (HDA), a sealed unit.

Magnetic tapes and flexible disks almost universally have a polyethylene terephthalate (PET) film substrate; common trade names are Mylar (DuPont), Celanar (Celanese), and Estar (Kodak). In a recent publication from the National Bureau of Standards it was concluded that, given storage at 20 to 25°C (68 to 77°F) and 50 percent RH, PET films are expected to have a lifetime of 1,000 years (Brown et al., 1984). Consequently, PET films will not be discussed further.

The magnetic particles used in most half-inch-wide computer tape are gamma-Fe_2O_3. In the most recent half-inch computer tape format (18-track NRZI, 19,000 BPI), CrO_2 is the magnetic material used. Most of today's video tapes use cobalt-surface-modified gamma-Fe_2O_3. Flexible computer disks use, in the main, gamma-Fe_2O_3, with increasing adoption of cobalt-surface-modified gamma-Fe_2O_3. It is believed that all these magnetic materials are stable chemical entities under normal storage conditions. All are produced by high-temperature (above 200°C) processes, which implies great stability around room temperature.

The magnetic stability of these particles is well understood. Their coercive forces are all above 300 Oe (more than 200 times the earth's magnetic field), and they are, accordingly, unaffected by the stray fields (about 10 Oe) associated with most electronic equipment. Their Curie temperatures (the temperature above which they become nonmagnetic) are above 400°C except in the case of CrO_2, which is only 120°C. These Curie temperatures are so far above the normal archival storage temperatures that no difficulty is anticipated. Other magnetic effects of concern include the print-through phenomenon, in which the magnetic fields from one layer of written tape can slightly magnetize the particles in the adjacent layers on the reel. The effect is known to be extremely small at room temperature but increases with temperature. However, at 65°C in a 4-hour test the print-through signal level typically remains a factor of 500 below the normal signal levels (Bertram and Eshel, 1979).

The binder systems in universal use in tapes today are of the polyester-urethane type. Because all tapes and flexible disks are intended to be operated with the writing and coding heads in as close physical contact as possible, the binder system has been chosen because of its extreme resistance to mechanical abrasion and its chemical stability. Most manufacturers of tape use slightly differing formulations, and no standards have yet been instituted.

The Achilles' heel of magnetic recording is the extremely close head-to-medium spacings required. Accordingly, most of the published reports deal directly or indirectly with the archival stability of the polyester-urethane binder systems. The particular area of concern is the hydrolysis of the binders. The basic reaction is (Brown et al., 1984; Cuddihy, 1980; Bertram and Cuddihy, 1982)

$$\text{Ester} + \text{Water} \rightarrow \text{Polycarboxylic Acid} + \text{Alcohol}$$

The problem with hydrolosis of the binder system is that its mechanical properties degrade if it is allowed to progress too far. It is thought that the adhesion of the binder to the substrate is particularly vulnerable to hydyrolysis (Brown et al., 1984). Earlier work showed that hydrolysis is a reversible reaction and suggested that not only can over-hydrolyzed tapes be recovered but that the system may equilibrate at a point where the hydrolysis reaction rate is zero (Cuddihy, 1980; Bertram and Cuddihy, 1982). To attain equilibrium at a satisfactorily low level of hydrolysis, tapes should be stored at 20°C (68°F) and 40 percent RH (Bertram and Eshel, 1979; Bertram and Cuddihy, 1982). Other satisfactory environments for limiting hydrolysis are shown in Figure 6-1.

A tape wound on a hub relies entirely on the maintenance of the layer-to-layer pressures and friction to transmit torque to the outer layers. The layer-to-layer pressure has been found to vary considerably when the tape reel is exposed to temperatures and humidities that differ from those that existed when the reel was

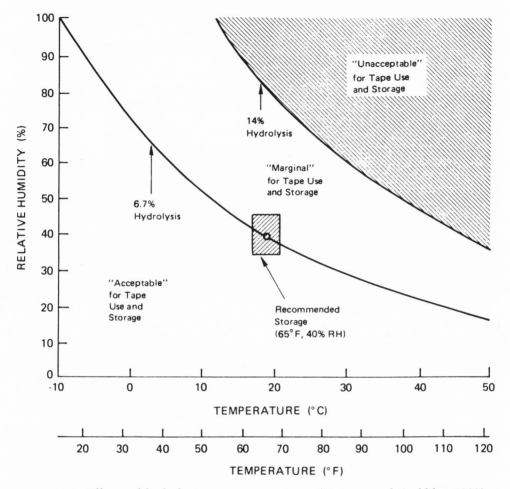

FIGURE 6-1 Effects of hydrolysis on magnetic tapes (Bertram and Cuddihy, 1982).

initially wound. This fact leads to the recommendation that archival tapes be stored at temperatures and humidities that are close to those that prevailed when the reel was wound. For this reason, the normal computer-room environment is selected for archival tape storage.

Moreover, the layer-to-layer pressure decreases as the tape tension relaxes. Mechanical creep in the PET substrate is such that over a period of about 5 years at 20°C the layer-to-layer pressures become insufficient to support high angular accelerations of the reel. When subjected to normal tape drive acceleration profiles, portions of the tape pack may then move with respect to each other ("cinching"). In addition, the tape is vulnerable to edge damage when careless handling subjects the reel to mechanical shocks. It is, therefore, advisable to rewind (retension) the reels periodically (Geller, 1983). The frequency of rewinding is clearly dependent on the accelerations anticipated. If all archival tapes are carefully rewound before being mounted on the computer tape transport, then very long storage periods are satisfactory. Indeed, many large computer tape archives do not include periodic rewinding in their tape maintenance programs (M. M. Cochran, private communication, 1985). After longer periods (say, 5 years) of storage, it is advisable to rewind the tape gently whenever it is withdrawn from the archive.

The published literature contains a host of other sound suggestions for the archival care of magnetic recording media. The review by Geller (1983) is particularly detailed and is highly recommended as a guide for action. Of the many excellent practices suggested, it is perhaps well to note those that concern the response of magnetic tape to changes in humidity. The coefficient of linear expansion of magnetic tape is almost the same for each percent change in relative humidity as it is for each degree Celsius change in temperature. Whereas the temperature may equilibrate in several hours, similar equilibriation of humidity in a tightly wound tape reel may take several days. It follows, then, that if an archival tape is to be subjected to a change in humidity (e.g., when brought to the computer room environment), it ideally should be allowed to equilibrate for several days even before being rewound or retensioned.

Given proper care, a safe conclusion can be drawn that a good-quality tape is an archivally reliable storage medium for periods of 10 to 20 years as a minimum, with much longer periods a distinct but as-yet unproved possibility. The essential point to be remembered is that today's recording media are most likely to outlive the period of utility of the other components of magnetic storage systems.

TRENDS

In the future the inexorable trend to higher storage densities will lead to magnetic tapes and disks that use very thin metallic layers (Mallinson, 1985a, 1985b). Projections are that not only will such tapes approach within a factor of two the storage densities of optical and magneto-optical disks, but also the physical properties of the metallic storage layers will be very similar. Thus, in magnetic disks, a 200-Å-thick layer of Co-Ni may be used, compared with magneto-optical disks that have a 150-Å-thick layer of Co-Fe-Tb and optical disks with a 150-Å-thick layer of a Te alloy. The archival properties of such metallic thin-film media, be they magnetic or optical, are not at present known; current estimates are for lifetimes of 10 to 30 years. Again, the salient point is that the recording medium may well outlast the hardware.

ARCHIVAL PROPERTIES OF SOFTWARE AND DOCUMENTATION

In a certain class of machine-readable records, namely those employing computer or digital technologies, a further archival problem arises. The mere recovery of the digital data is not possible without some software. The proper operating system (called software to distinguish it from hardware) must be available at the time that data are to be recovered. Sadly for the archivist, software today is changing more rapidly than hardware; for example, Western Electric's UNIX operating system has been offered in about 30 versions over the past decade. Offsetting this serious problem are, of course, some potentially attractive reasons for using digital recording techniques; compatibility with the computer environment and the ability to perform perfect error detection and correction are prime examples.

Given the proper operating system for reading out the digital data, a further requirement arises. Appropriate documentation must be at hand that will provide

Controlled-environment tape storage area. Archival care of magnetic recording media requires attention to conditions of temperature and humidity as well as careful periodic rewinding.

the necessary information on the digital codes used, the organization or format of the record, and several other minor but critical details. Operating systems are usually resident on computer tapes or floppy disks, thus compounding an already difficult archival problem. The documentation may be in machine-readable or human-readable form but, given the human species' well-known tendency to procrastinate, the needed data may well be incomplete or missing (National Academy of Sciences, 1982).

ARCHIVAL PROPERTIES OF HARDWARE

The fact that most electronic hardware is expected to function for no more than 10 to 20 years raises very serious problems for long-term (more than 20 years) archival preservation. Even if the operating systems and documentation problems somehow are dealt with, what is the archivist to do when the machine manufacturer declares the hardware obsolete or simply goes out of business? Will there be an IBM or a Sony in the year 2200? If they still exist, will they maintain a 1980–1990 vintage machine? Moreover, it must be realized that no archival organization can hope realistically to maintain such hardware itself. Integrated circuits, thin film heads, and laser diodes cannot be repaired today, nor can they be readily fabricated, except in multimillion-dollar factories.

The inescapable conclusion is that, if a long-term archive preserves records in machine-readable form, it will be committed eternally to file conversion (i.e., rerecording the old obsolete versions into the new current format) approximately every 10 to 20 years. Not only would such an operation be enormously expensive, but also, in an archive—where by definition no records can be disposed of—it is a task that grows exponentially with time. Precisely such file conversions take place all the time, of course, in today's computer facilities, but the critical difference is that records in such facilities are continually being retired—perhaps to be sent to an archive!

CONCLUSIONS

The committee's conclusions in the area of magnetic media are as follows:

1. Magnetic recording media today are of sufficient stability that only short-term (10 to 20 years) storage is practical.

2. Operation of short-term magnetic tape archives in accordance with the recommended storage practice detailed by Geller (1983) is possible.

3. Magnetic recording media and other machine-readable recording media cannot be recommended for long-term (over 20 years) storage because of the difficulties in maintaining software, hardware, and documentation; provision for repeated file conversion can overcome this limitation.

REFERENCES

Bertram, H. N., and E. F. Cuddihy. 1982. Kinetics of the humid aging of magnetic recording tape. IEEE Trans. Magn., 18(5, September):993–999.

Bertram, H. N., and A. Eshel. 1979. Recording Media Archival Attributes (Magnetic). U.S. Air Force Systems Command, RADC F 30602:78:C-0181.

Brown, D. W., R. F. Lowry, and L. E. Smith. 1984. Predictions of Long-Term Stability of Polyester-Based Recording Media. National Bureau of Standards, NBSIR 84-2988, December. See also Kinetics of hydrolytic aging of polyester urethane elastomers, Macromolecules, 13:248-252 (1980); Hydrolytic degradation of polyester polyurethanes containing carbodiimides, Macromolecules, 15:453-485 (1982); Equilibrium acid concentrations in hydrolyzed polyesters and polyester-polyurethane elastomers, J. Appl. Polym. Sci., 28:3779-3792 (1983); and Hydrolysis of crosslinked polyester polyurethanes, Div. Polym., Mater. Sci. Eng., 51:155-161 (1984).

Cuddihy, E. F. 1980. Aging of magnetic recording tape. IEEE Trans. Magn., 16(4, July):558-568.

Geller, S. B. 1983. Care and Handling of Computer Magnetic Storage Media. National Bureau of Standards, NBS SP 500-101, June.

Mallinson, J. C. 1985a. The next decade in magnetic recording. IEEE Trans. Magn., 21 (3, May):1217-1220.

Mallinson, J. C. 1985b. Archiving human and machine readable records for the millenia. Society of Photographic Scientists and Engineers Second International Symposium: The Stability and Preservation of Photographic Images. Ottawa, Canada. August 1985.

National Academy of Sciences. 1982. Data Management and Computation, Vol. 1: Issues and Recommendations. Washington, D.C.: National Academy Press.

National Archives and Records Service. 1984. Advisory Committee on Preservation White Paper: Strategic Technology Considerations Relative to the Preservation and Storage of Human and Machine Readable Records. July. Unpublished.

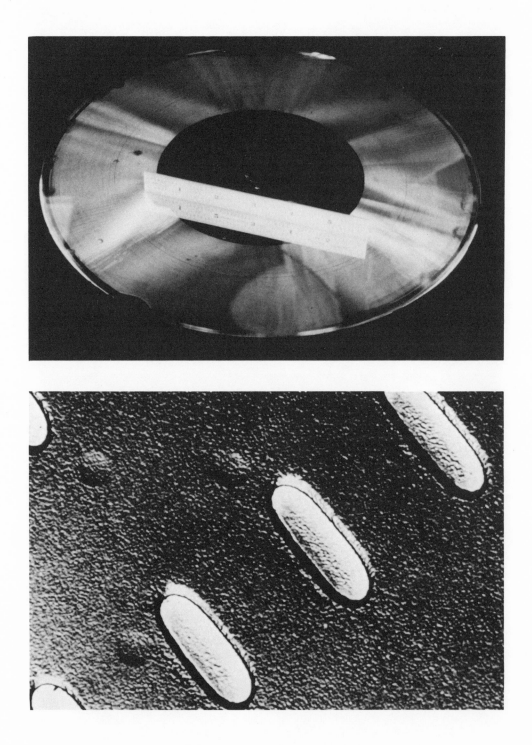

Optical disk—a recent development in data storage technology. The magnified view shows digitized data encoded on the disk.

7

Optical Disks

The optical disk is the newest of the storage media, and because of its newness a great amount of confusion exists in terminology, technology, and standards. Many varieties of optical media exist, and it is futile to attempt to make definitive statements as to archival life at the present time. The most positive statement about archival life is that there exists a general belief that many forms of optical disk media will retain data for 10 to 20 years. A look at the natural progression of data systems of recent technology indicates that the useful system life is only about 10 to 20 years. Therefore it is not reasonable to ask that data storage media exist longer than the system capable of playing back the data. The implication is clear. If data are stored on optical disk, the data must be rerecorded in 10 to 20 years to ensure their existence. As a result, the cost and time to rerecord must be considered as a necessary expense of using optical disk as a storage medium. General agreement exists at present that no direct evidence is available to substantiate a claim for more than 20 years' life for data recorded on an optical disk (Joseph W. Price, Library of Congress; John C. Davis, National Security Agency; and David H. Davies, 3M Company, presentations to the committee, 1985; see also Price, 1984; Davies, 1985; Davis, 1982).

Optical disk technology is an outgrowth of the home entertainment industry. The early work on optical disk, called videodisk (Table 7-1) in the late 1960s and early 1970s, was directed at producing an inexpensive "pirate-proof" alternative to prerecorded video tapes. The videodisk is the video analog of audio records, since both contain prerecorded information. Videodisks are produced by first using a laser to write on the photosensitive material on a glass master, and then a metal master is made by using injection molding. The video information is impressed on plastic disks, similar to the process used for stamping out high-fidelity records. The videodisk has met with only limited market success because of the availability of low-cost video tape recorders.

The latest disk entry in the prerecorded domain is called the compact disk (CD) or the digital audio disk. The CD, which stores ultra-high-fidelity digital audio on a 4.7-in. disk, appears to be meeting with exceptional market acceptance. This acceptance is seen as helping provide a continuing technology base that can generate further applications interest for all the varieties of optical disks.

Early in the development of the videodisk, many in and outside of the computer industry saw in it the promise for much higher storage densities than with the magnetic disk, and research was initiated into the technology for its use in data storage. The results of this research now are beginning to appear as products in the marketplace. The compact disk read-only memory (CD ROM) plays back computer-compatible prerecorded digital data. An optical digital disk for more general computer application is "write-once," wherein the data written by the user cannot be erased. This differs from CD and CD ROM, which are publishing media prerecorded as a final value-added step in manufacture.

The next significant event for optical disk will be the introduction of erasable optical disks, which are expected to be available within the next 2 to 3 years (Goldberg, 1984). The erasable optical disk will use a magneto-optical or phase change material for which no estimate of archival life can be made at the present time. When the erasable optical disk appears, the competition between optical and magnetic recording will intensify. The removability (from the disk drive) of the optical media disk will offer a significant advantage in some applications.

PERMANENCE OF OPTICAL DISK MEDIA

The optical disk is a new technology with new media processes, new read-write mechanisms, new data analysis methods, and many new companies starting in the business (Fujitani, 1984). There are four basic markets for optical disks: (a) prerecorded entertainment programs, (b) interactive education and training, (c) document storage, and (d) computer digital data storage. Of these markets, only a very small portion of the document storage and data storage market requires the archival quality desired for 50- to 100-year life. If the optical disk turns out to be archival, it will be a fortuitous result and not the result of market forces.

A wide range of application formats is available today (Table 7-1), and in 2 years erasable entries can be added to the Read or Write column.

An examination of write-once media options available today (Table 7-2) shows that 194,400 possible combinations exist, not all of which affect archival life. Of these possibilities, 10 to 20 working write-once systems have been built;

TABLE 7-1 Applications Formats for Optical Disk

Read or Write	Data Format	Common Names
Read only	Analog video	Videodisk
Read only	Analog data	Digital videodisk
Read only	Digital audio	Compact disk (CD)
Read only	Digital data	Compact disk read only memory (CD ROM)
Write once	Video/image	—
Write once	Digital data	Optical digital data disk (OD³)

NOTE: Erasable optical disk anticipated in 2 years.

TABLE 7-2 Write-Once Options for Optical Disk (Select One from Each Column)

Substrate	Active Layer	Mechanism	Construction	Size (in.)
Glass	Monolayer	Bubble former	Air sandwich	14
Aluminum	Bilayer	Hole former	Sealed laminate	12
Plastic	Trilayer	Alloy former	Core sandwich	8
	Quadrilayer	Phase change	Flat sheet	5.25
		Dye ablate		4.7
		Dye bleach		

Protection	Tracking	Detection	Format	Encoding
None	Nongrooved	Absorption change	Preformat	One for each
In contact	Grooved track	Reflectance change	Nonformat (user)	drive system
Off contact	Intermittent track	Optical phase change	Postformat	
	Push-pull			
	Wobble			

SOURCE: 3M Company (David H. Davies).

however, a lack of uniformity among these diverse systems precludes any definitive statement as to archival life. A market consolidation must occur within the next few years because this wide diversity is technically interesting but not commercially viable. No surprises are expected in the useful life of the write-once disk, which manufacturers project at 10 to 20 years. It will be at least 1 or 2 years before more definitive data are available on its permanence.

STANDARDS

Although the need for accepted standards is widely recognized, no official standards are available. The industry has made attempts to start the standards process, as shown by the workshop held in June 1983 on Standardization Issues for Optical Digital Data Disk Technology (National Bureau of Standards/National Security Agency, 1983). In addition, the American National Standards Institute (ANSI) formed committee X3B11 on Optical Digital Data Disk in 1984 to formulate standards. When standards have been established for the write-once formats, archival testing should produce some useful data.

The read-only media have existed in a stable format for 10 years. A disk is manufactured from a mastering process in which the format, software, and data can be replicated for mass distribution. Read-only disks are not written serially as in the write-once system but are stamped out singly as a full disk, and the cost per disk is determined by the size of the production run. Sufficient field experience exists for read-only disks to allow an extrapolation of a useful life of 10 to 20 years.

PRESERVATION, USE, AND STORAGE

The factors of storage that are most harmful to the optical disk are heat and humidity. Humidity causes oxidation of the recording surface, and heat accelerates the process. Some of the other factors that cause concern are the adhesion of various layers one to another, catalytic corrosion, galvanic corrosion, and mechanical stresses. The expected storage conditions for optical disks are typical of a computer room environment with controlled heat and humidity (nominal 20°C, 45 percent RH).

In use the optical disk suffers no measurable degradation from continuous reading of the same data, since a much-reduced power (approximately 10 percent of write power) laser beam is used and the reading head has no physical contact with the surface.

TRENDS AND PROJECTIONS

Large on-line digital data storage systems have existed within the government since the 1960s. Examples are listed in Table 7-3. These data storage systems have a maximum useful life of 10 to 20 years, after which the system is no longer maintainable because parts and service are difficult to obtain and the system figuratively crumbles to dust. The important point of this revelation is that the data may still exist on the media but the equipment to retrieve the data does not exist. The obvious solution is to rerecord the data on a new system before the old system collapses. Here cost becomes an important factor. This situation raises a fundamental question regarding future actions: Is the National Archives ready to become a data processing facility with all its attendant problems?

The trend within the next 3 to 5 years for write-once and erasable optical disk systems is toward standardization. Not all existing and proposed formats can be sustained by the marketplace. The particular formats that emerge are not critical; the critical factor is the reduction of market instability. When market stability occurs, the optical disk will begin to approach the maturity of magnetic recording and will begin to validate the promise of 10 to 20 years of useful system life.

The projection for storage density, shown in Figure 7-1, indicates that the optical disk will have the density advantage for at least the next 15 years. After that time, the issue will not be storage density but rather system issues such as removability of the media, system reliability, and, most important, cost. Low cost demanded by the consumer market and archival life for long-term data storage are diametrically opposed goals. Only peripheral considerations will be given to the archival qualities of the optical disk.

TABLE 7-3 Examples of Typical Digital Data Storage Systems

System	Time	Bits
IBM Harvest Tractor Tape System	1962–1976	3×10^{11}
IBM Photo Digital Store	1970–1972	2×10^{12}
Ampex Terrabit Memory (TBM)	1972–1982	2×10^{12}
Bragen Automated Tape Library (ATL)	1975–	10^{12}
IBM 3850 Magnetic Tape Cartridge System	1977–	10^{12}

SOURCE: National Security Agency (John C. Davis).

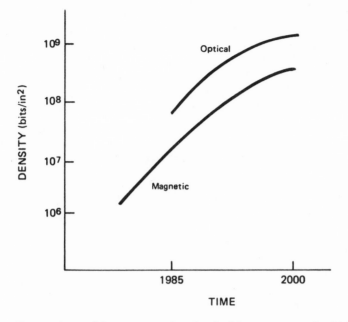

FIGURE 7-1 Comparison of data storage density (in bits per square inch) for optical and magnetic systems, projected to the year 2000.

ADVANTAGES, DISADVANTAGES, AND CONCLUSIONS

Advantages

The advantages of optical disk are as follows:

1. The data are in digital form for ease of manipulation.
2. The data are on a disk for rapid access.
3. The data can be read indefinitely (i.e., without degradation due to reading).
4. The data are in a nonerasable form.
5. The data are not affected by magnetic fields.
6. The cost to store the data is low.

Disadvantages

The disadvantages are these:

1. The data are not human readable.
2. The data cannot be manipulated without the software and hardware used to write the data.
3. The data require recopying at least every 20 years.
4. Separate software documentation for data must be maintained.
5. The next generation of new recording hardware is required at least every 20 years.

Conclusions

At present, it cannot be proved that the optical disk is archival in either the read-only or the write-once format. The consensus of the optical disk industry is that a useful life of 10 to 20 years can be reasonably assured for the read-only disk and extrapolated to 10 to 20 years for the write-once disk. No hard evidence can be offered for the useful service life of the hardware or archivability of the write-once disk in light of the volatility of the market. If one chooses to use the optical disk, knowledge available today on current optical disk media and systems indicates that the data must be recopied at least every 20 years.

REFERENCES

Davies, D. 1985. Optical Media Life Considerations. Optical Record. 3M Co., unpublished.
Davis, J. C. 1982. Mass storage systems—A current analysis. Digest of Papers From the Fifth IEEE Symposium on Mass Storage Systems, October 26–28.
Fujitani, L. 1984. Laser Optical Disk, The Coming Revolution in On Line Storage. Communications of the ACM, 27(6):546.
Goldberg, M. 1984. Optical disk hardware performance and availability. Digest of Papers From the Sixth IEEE Symposium on Mass Storage Systems, June 4–7.
National Bureau of Standards/National Security Agency. 1983. Proceedings of Workshop on Standardization Issues for Optical Digital Data Disk Technology. National Bureau of Standards Special Publication 500-111.
Price, J. 1984. Videodisc and Optical Disk. The Optical Disk Pilot Program at the Library of Congress. November–December. Unpublished.

SUPPLEMENTARY BIBLIOGRAPHY

American Council of Learned Societies. 1985. Committee on the Records of Government. Washington, D.C. March.
Council on Library Resources. 1985. Videodisc and Optical Disk Technologies and Their Application to Libraries. Washington, D.C.
National Archives and Records Service. 1984. Strategic Technology Considerations Relative to the Preservation and Storage of Human and Machine Readable Records. Prepared by Subcommittee C of the NARS Advisory Committee on Preservation, July 1984. Unpublished.

Microfilm reading room at the National Archives. Researchers use microfilm records in lieu of original documents.

Researcher examining document at the National Archives. Condition of the document illustrates the need for preservation.

8

Discussion of Findings

This discussion considers the facts that the committee has gathered concerning the deterioration of the collections of the National Archives and the conditions that have contributed to this deterioration. The committee has adopted some general recommendations, and these are discussed first. The second section concerns the possibility of mass treatment for NARA's holdings, in particular environmental controls and deacidification, and there are recommendations in those areas. Finally, there is a discussion of the coordinated steps the Archives can take to determine when preservation action is advisable and which methods of preservation are appropriate.

NATURE OF THE PROBLEM

The National Archives, founded in 1934, is a young institution. Although some of its records are more than 200 years old, the collection is not old in terms of the history of paper records or of paper itself, for some papers have lasted many hundreds of years. The bulk of the collection was created during the past 100 years, however, and therefore has accumulated the problems associated with the decline in purity of western papermaking that began in the 17th century. This decline accelerated in the mid-19th century when manufacturers, responding to ever-greater demand, began to produce cheaper papers dangerously prone to oxidative and hydrolytic reactions. In the mid-20th century, copying processes were developed that used paper that deteriorates even more quickly and, in some cases, suffers relatively rapid image loss. As these sorts of records have begun to reach the limits of their durability, the problems of failing paper and vanishing image are becoming of concern to the world's libraries and archives. This concern is certainly appropriate, but dire predictions about the limited useful life expectancy of 19th- and 20th-century papers have all too often led to panic and a sense of helplessness, sometimes producing hasty, unthinking responses rather than the realistic, planned course of action that is needed.

Condition of the Collection

The collections of the National Archives contain a great range of different kinds of damaged or deteriorating records. Inspection by members of the committee revealed that papers seem to be in worse condition at the beginning or end of a group or along the edges (that is, where they are most handled or most exposed), whereas the condition in the center of a group seems to be good. The sorts of damage are what would be expected: tears, discoloration, loss of fold endurance, and image loss. The volume of large collections of damaged or deteriorating paper as a percentage of the total volume needing remedial action is small. Most of the records that need treatment occur as single sheets or in small groups.

NARA reported the following points of concern:

- A large percentage of the collection is quite acidic, with a pH less than 5.
- 160 million sheets have already suffered major damage of one form or another; a very small percentage of records cannot be handled at the present time.
- 270 million sheets were produced by quick copy reproduction processes such as Thermofax and Mimeograph during the 1940s, 1950s, and 1960s, and these are rapidly deteriorating; loss of information in the earliest of these records is expected within 10 years.
- 100 million sheets are subject to mechanical damage by frequent handling.
- The estimated total number of these records at some risk of loss of information value is 530 million, or over one-sixth of the entire collection.

Prevention of Future Deterioration

Besides environmental controls, which are the subject of a separate section, some improvements can be made in current practice that would help to prevent future deterioration of the sort now evident in NARA's collections. In particular, the adoption of media standards where they exist, and their development where they do not, would increase the future volume of stable records, while less frequent handling would increase the lifetime of all NARA's records.

Paper and Film

The compositional factors that most influence the durability of a paper are the quality of the fiber and the level of acidity. The standards in these areas are well known, as are the aspects of the manufacturing processes that affect achievement of the standards. These standards are given in detail in Chapter 4. It is only recently, however, that there has been an economic advantage to manufacturing, and hence to purchasing, permanent papers. Papers of archival permanence are now available at a cost reasonable enough to encourage the federal government to use such paper for a broader range of documents.

There also exist standards for photographic film in the areas of film type, processing, and storage conditions. These standards, adopted by the American National Standards Institute, are given in detail in Chapter 5, and the National Archives should apply them as a matter of course, while keeping in mind the need to maintain good resolution, contrast, framing, and completeness.

Magnetic Tape and Optical Disk

In the areas of magnetic and optical recording media, the committee found that there are three problems that prevent records of these types being considered archival. The first is the pace of technological development, which has the result that changes in media and hardware require periodic conversion of older records into newer formats. The second problem is that the life expectancy for magnetic tapes is estimated to be 10 to 20 years, and a similar lifetime is projected for optical disks. The third concerns the lack of standards in certain areas, which results in incompatibility of systems, unpredictability of product lifetime, and uncertainty in planning.

Chapters 6 and 7 point out two difficulties that result from the pace of development. First, rapid development requires systems conversion, and second, there are the attendant problems of the future availability of older hardware and the preservation of software and documentation necessary to do the conversion. Neither of these problems has an obvious solution, but study should be undertaken now to prepare for the future difficulties that will result from the inevitable deposit of large volumes of machine-readable records with NARA.

The life expectancy of the media also makes periodic recopying necessary. A medium might be archival and yet require recopying, but the frequency of recopying could be less frequent than the 10 to 20 years that magnetic and optical media now require. In addition, deterioration in these media is not easily perceived, which means that even more frequent reading would be necessary. Data compaction in this case is a two-edged sword because, although it allows processing large amounts of data, even small-scale physical deterioration in the media will have a large-scale impact on the condition of the information.

Chapters 6 and 7 also point out that there are no standards for the binder systems used in manufacturing magnetic tape and that there are no standards for optical disks in the areas of format, hardware, and materials. The fact that significant recording on these media is now being done in the federal government suggests that work on standards needs to be speeded up significantly and that it is important for the government to take a lead in establishing such standards. The implications of this concern for preservation action by NARA are discussed later in this chapter.

Handling

It has been shown that the rate of natural aging of paper that has been subjected to improved environmental conditions is not a linear phenomenon. Paper that has reached a point of zero folds may last indefinitely as an integral sheet if left relatively undisturbed and in a suitable container in controlled conditions. It is extremely rare to find disintegrated paper fragments in a box of records without a history of use.

Unfortunately, there is no current standard for determining the difference between frequently and infrequently used archival documents. The section, A Systems Approach, later in this chapter contains a further discussion of this question. No matter what the definition for frequency of use, however, less use means longer life for paper records. It is possible that certain additional finding aids may reduce use in selected groups of records, and it might be useful for NARA to investigate the effectiveness of new finding aids in preservation. NARA should also encourage better indexing by the depositing agencies. Nevertheless, the cost of comprehensive indexing is so high that it cannot be recommended simply as a preservation strategy.

Apart from deterioration of media, however, handling may have other negative consequences. As discussed in the description later in this chapter of the decision procedure for preservation action recommended by the committee, one of the criteria that affects preservation strategy is the cohesiveness of the record. There, the subject is the difference in verification of copying appropriate to files that contain redundant information as opposed to files whose individual documents hold unique places in the record. In the latter cases, the loss of a single document, or perhaps even part of a document, may have a significant effect on the information value of the file. Frequent handling of such files increases the risk of loss, even when the medium itself is in good condition and is expected to remain in good condition in the future. Copying at the discretion of the archivist may be justified in these cases.

General Recommendations

The committee feels that several recommendations covering issues of general concern are appropriate:

• A general improvement in the quality of paper used by the federal government would be an important step in minimizing future problems of the sort now experienced by the National Archives. Since permanent papers are becoming available at a reasonable cost, the government should use such papers for records that have permanent value.

• Archival standards are available for papers and photographic films. Archival standards are also available for electrophotographic reproduction. NARA should ensure that the records it creates or copies with these media or processes meet these standards.

• Archival standards do not exist for magnetic tape or optical disk or for the reproduction of records on such media. Since these media are currently being used by the federal government, and since their use will greatly expand in the future, NARA should promote the development of standards for these media at the earliest possible date.

• Major deposits of machine-readable records exist. If these records are to be useful to future research at the National Archives, NARA should be prepared to accession them and to preserve the information they contain.

OPTIONS FOR MASS TREATMENT

The difficulties associated with the distribution of damaged papers in the NARA collections were mentioned previously, and this subject is treated in greater detail later in this chapter. There still remains the question as to whether large-scale treatment might be appropriate for the records as a whole. The committee looked at two aspects of this problem—environmental control and mass deacidification.

Environmental Control

The National Archives has been relatively forward-looking on environmental controls; air conditioning was installed in the 1930s. In more recent years such concerns have expanded to include areas beyond temperature and humidity. Furthermore, while papers in the Archives building are comparatively well housed, their histories are spotty at best. Materials that are stored for years in agency files, and then for more years in various federal records centers, are subject to frequent changes in temperature and humidity and to little if any controlled protection against pollutants. These records arrive at NARA carrying their history with them.

Wide fluctuations in temperature and humidity, especially familiar in such cities as Washington, are a major contributor to the rate at which modern papers deteriorate. Establishing the rate of deterioration in records is a key requirement for planning preservation strategies. Unfortunately, few scientific data exist for determining such rates. However, it is certain that moving into the Archives Building was salutary for the vast majority of NARA's 3 billion records and that improved conditions have slowed the rate of deterioration. The improved environmental controls that NARA is now planning to implement will slow the rate of deterioration even further. Although questions remain concerning the precise standards for environmental controls, the committee feels confident in endorsing the standards recommended in Chapter 3.

The committee notes that the critical environment is that at the surface of the paper and that the greater part of NARA's collection is boxed. Unfortunately, very little is known about the microenvironment of the archival storage container. Indeed, what is known suggests that good containers may obviate the need for the most exacting and costly macroenvironmental control; on the other hand, it is also known that the close proximity of acidic materials is undesirable. The committee feels that a substantial research effort should be conducted in this area, especially as NARA is planning to install compact shelving that will further encase the records and may significantly influence the effectiveness of any improved macroenvironmental control system.

Mass Deacidification

A high percentage of NARA's 3 billion records is categorized as quite acidic, with a pH of less than 5. While this condition can affect the physical strength of documents, it does not ordinarily result in the destruction of documents on its own (acidic inks are another matter). Deacidification, however, cannot restore the strength of damaged paper, and the process itself could result in further deterioration. The point of a mass deacidification program is to treat the papers in bulk, but this is not an option for the National Archives because its records present a great mix of different types of documents that could not economically be separated. The deacidification process would result, for example, in heat damage to Thermofax documents and solvent damage to some inks. On the other hand, there could be an added benefit in deacidification in the future for collections of records that have good physical strength properties. The cost of using the process as a preventive, however, is not justified at this time.

Recommendations on Mass Treatment

The committee developed the following recommendations regarding mass treatment:

• The standards given in Chapter 3 for temperature, humidity, and pollutants should be implemented (see Tables 3-4 and 3-5 for specific standards).
• NARA should conduct a study of archival storage containers and microenvironments, including boxes, folders, and polyester encapsulation, with a view to understanding the maximum benefit that can be obtained from particular materials and designs. The committee feels that this is an underexplored area that may yield results highly significant to NARA's preservation efforts.
• NARA should not undertake a mass deacidification program at this time but should monitor the development of deacidification processes.

A SYSTEMS APPROACH

This section describes a general procedure for preservation action—in effect, a decision tree for choosing which records need treatment and what treatment to give. The first part of the discussion concerns the media that are appropriate to archival copying. The next section describes that procedure itself, together with its limitations and subcategories. Finally, there is a discussion of implementation.

Suitability of Preservation Copying Media

The holdings of the National Archives are predominantly paper-based and will remain so for the foreseeable future. The question remains, however, of what to do with records at risk. The committee examined alternative media onto which records might be copied:

• *Copying onto magnetic or optical media*, which would have the advantage of compaction and potential ease of transmission and future manipulability;
• *Copying onto paper*, since archival standards for paper and for electrophotographic copying exist; and
• *Copying onto microfilm*, which offers the advantage of compaction and for which archival standards for both film and microphotographic processes exist.

Copying Onto Magnetic or Optical Media

Although they are quite useful and stable for time periods of 10 to 20 years, current flexible magnetic recording media suffer from recognized material degradation processes that make them vulnerable to large-scale information losses over long periods of time. Their advantages of rapid access times and adaptability to large-scale data manipulation are not particularly relevant to archival needs. Their

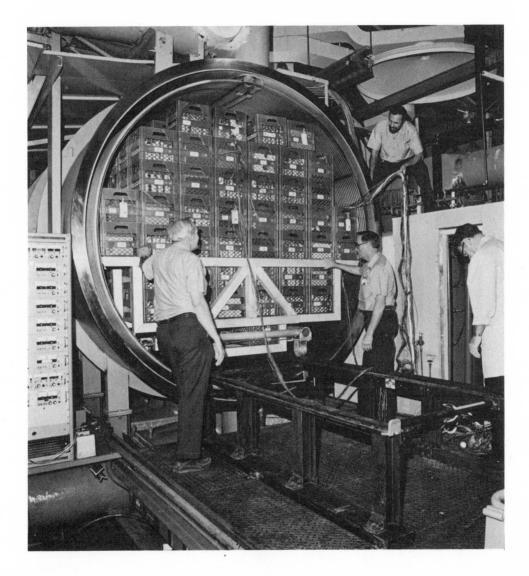

Chamber for mass deacidification of books. The deacidification process can add years to the life of endangered books but is not now practical for preservation of diversified collections of individual documents.

ability to be erased or updated is, in fact, a liability in archives. Material degradation processes in current optical disks are not well understood, and the technology is not sufficiently developed to recommend them for archival use.

Both magnetic tape and optical disk storage are vulnerable to the problems of hardware obsolescence and software and documentation loss discussed earlier that could render data unreadable even if it still remained preserved on the primary medium. This is a serious concern for archives that must plan for the ages, as opposed to libraries that must serve a variety of users primarily interested in rapid access and retrieval.

Copying Onto Paper or Microfilm

Standards for archival-quality paper and electrophotographic duplicating processes have been developed that are adequate for the National Archives to use in establishing and maintaining quality control for copying damaged documents onto paper. In addition, the long history of generally satisfactory use and the ability of current holders of large silver-based microfilm collections to maintain adequate quality control make this medium archivally acceptable. It is possible to transfer records from paper or film to magnetic or optical storage media using automated systems, so that conversion in the future should not be labor-intensive.

Recommendations on Archival Copying

The committee recommends the following with regard to copying media:

• The media that are appropriate for archival preservation are paper and photographic film, and the processes appropriate to copying using these media are archivally standard electrophotographic processes (for paper) and silver-based micrographic processes (for film).

• The materials and technical problems inherent in the use of magnetic and optical storage media and the lack of suitable standards for archival quality make their use as preservation media for archival storage inappropriate at the present time.

Decision Tree for Preservation Action

The committee has organized its recommendations for action into the decision tree shown in Figure 8-1. Here, the collection is described primarily by the frequency of use and the physical condition of the documents. The definitions of use and condition are discussed first, followed by paragraphs giving the committee's recommendations for each subcategory.

Preliminary Considerations

It seemed initially that the question posed by the National Archives was purely practical: What recording medium is most appropriate for the long-term preservation of the nation's records? It became clear, however, that an answer to this question was not possible merely on the basis of facts about materials science

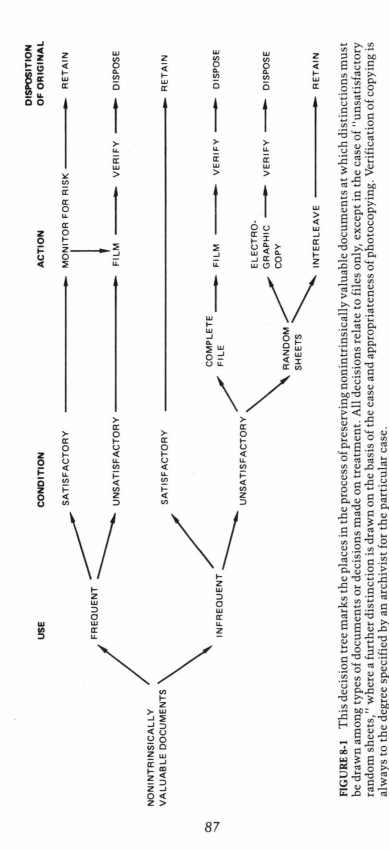

FIGURE 8-1 This decision tree marks the places in the process of preserving nonintrinsically valuable documents at which distinctions must be drawn among types of documents or decisions made on treatment. All decisions relate to files only, except in the case of "unsatisfactory random sheets," where a further distinction is drawn on the basis of the ease and appropriateness of photocopying. Verification of copying is always to the degree specified by an archivist for the particular case.

alone. The original question requires consideration of problems such as media stability and lifetime, reproducibility, cost, etc., but the committee discovered that determining the appropriate application of a particular medium frequently requires decisions based on archival criteria. The National Archives must clarify the key criteria of *use* and *condition* and must improve the statistical description of its collection before remedial action can be taken to improve the condition of damaged records.

Criteria for Use   Although frequency of use here must be defined within an archival context, if it is to be employed to justify preservation action it still must correspond to a reasonable concept of use. The average document in the National Archives is seen perhaps only once every 100 to 200 years. Frequent use might then be defined as anything more than twice a century, but this level of use does not justify preservation action. As another example, there are popular series made up of case files with good indices whose individual documents are not frequently handled because the nature of the file and the depth of the indexing lead the researcher directly to the desired papers in most cases (Schofer, 1986). On the other hand, there are correspondence series of great historical interest that are examined document-by-document by several researchers in a year. It is these latter series that qualify for preservation action because of frequency of use. NARA does not have a well-defined basis for clearly identifying documents that are frequently used in this sense. NARA must develop such criteria to establish priorities for copying.

Criteria for Condition   Documents in satisfactory condition are those that are reasonably expected to retain their information value for the foreseeable future, whereas documents in unsatisfactory condition are those that are at risk of loss of their information value even without being able to say precisely when. At the moment, NARA cannot easily distinguish between these two categories to allocate preservation resources and to adopt reasonable preservation schedules. NARA will need a specific guideline for life expectancy, say 100 years, that should be related to the time before which the document will be seen again. Given this time period, an archivist should be able to say of any document or series whether it will retain its information value at the end of that period. If it will not, it should be scheduled for preservation. NARA will also need criteria that will allow it to say whether one kind of damage is more serious than another, so that faster or more extensive treatment can be scheduled.

Problem Distribution   A potential difficulty in any large-scale preservation effort was pointed out earlier: The majority of the records at risk of deterioration exist as isolated documents distributed at random among the National Archives holdings. Only a much smaller fraction exists as identifiable series of records. Specific data on the number of documents in these categories would be valuable in constructing a general plan.

The type of information NARA needs can be obtained only by actual assessment of the conditions of individual records series. It is obvious, of course, that NARA cannot perform an item-by-item, or even container-by-container, survey of its collection. There are procedures, however, that would improve the information available for preservation planning. First, records must be kept of the condition of papers that actually are seen or used by staff and patrons, much as in the experimental intercept program described by Calmes et al. (1985). In such a program documents should be screened for their condition before being given to the person requesting them, and records should be kept of that condition. Second, the Archives is currently designing a program to improve the maintenance of its holdings, which may result in the construction of a shelflist. It may be possible to generate statistical information about the collection from such a shelflist. Exploration of this possibility should be included in the planning of the program.

Recommendations for Preservation Action

The committee recommends the following as preservation actions:

• The National Archives should institute procedures immediately that will yield statistics concerning damaged records that are useful for long-term preservation planning and for deciding treatment priorities. The committee suggests that statistics be kept that reflect the condition of records used both in the reading rooms and by the staff and that these statistics be supplemented by the information generated by the ongoing maintenance operations.

• The National Archives should establish criteria for frequent and infrequent use and for satisfactory and unsatisfactory conditions so that priorities in treatment may be assigned.

Preservation Actions by Categories

Frequently Used, Satisfactory Condition   Frequently used files, although currently in acceptable condition, are at risk simply because of their use. This is based on the assumption that the most frequently used documents contain information of greater-than-average value and will continue in high use in the future. The increased risk for these frequently used materials justifies monitoring to allow prompt action should their condition change. Monitoring probably can be conducted most efficiently at the time of use and could make use of researchers in identifying marginal condition. As increasing numbers of documents within a file are identified as damaged, the possibility of filming the entire file should be considered.

Frequently Used, Unsatisfactory Condition   Frequently used files that are in unsatisfactory condition demand timely preservation that can be accomplished most readily by microfilming. The greater the frequency of use, the lower the degree of risk acceptable before action is taken. For all microfilming, the verification of the images should be done at a level specified by an archivist, preferably with the advice of a committee of users who can help determine the degree of loss if some small fraction of the images is lost. Many records contain much redundant information, so little risk of real harm would be associated with the loss of a single image. Other records are valuable because they are unique sources of certain information, and thus loss of even part of a single image would cause serious harm. After appropriate verification, NARA would dispose of the paper documents.

Infrequently Used, Satisfactory Condition   Documents that are seldom used and in satisfactory condition obviously need no particular attention other than being included in the continuing collection of statistics on use and condition to better characterize the holdings.

Infrequently Used, Unsatisfactory Condition   For those files classified as infrequently used and also in unsatisfactory condition, a further distinction should be made on the basis of whether the damaged documents occur as isolated sheets or as large groups of similarly poor-condition papers. Single damaged sheets should be electrophotographically copied using materials and processes that conform to archival standards, and NARA may dispose of the originals. This can be done on a continuing basis as damaged sheets are identified by users. Damaged sheets that contain notations in colored pencils, different inks, or watermarks may not retain these distinctions after copying. These documents may be preserved by interleaving between sheets of polyester film.

Extensive files of damaged or poor-quality documents, such as mimeograph copies, should be microfilmed. The level of verification should again be determined by an archivist but may be done to a different standard than in the case of frequently used documents. NARA should dispose of the original documents when verification is complete.

Implementation

Implementation of the decision tree will require that the National Archives commit itself to a very intensive program of quality control and verification of copies. Each step in the electrophotographic or micrographic process will need rigid standards of performance, and the work produced under these standards will need continual inspection. In some cases, especially those in which infrequently used materials are being filmed in series, it may be sufficient to verify the photographic quality while only sampling for content verification. On the other hand, frequently used materials may need to be verified frame-by-frame.

Recommendations on Preservation Strategy

The committee recommends implementation of the following preservation strategy:

• The National Archives should adopt the decision procedure and the recommendations on treatment and records disposal that are embodied in the decision tree, Figure 8-1, with the caution that this recommendation cannot be separated from the following one.

• Portions of the proposed preservation plan include disposal of original records after copying. In these cases, the copy *will be* the record. The National Archives must establish in perpetuity a program of effective quality control and verification of copying.

REFERENCES

Calmes, A., R. Schofer, and K. R. Eberhardt. 1985. National Archives and Records Service (NARS) Twenty Year Preservation Plan. NBSIR 85-2999. Gaithersburg, Maryland: U.S. Department of Commerce.

Schofer, R. E. 1986. Cost Comparison of Selected Alternatives for Preserving Historic Pension Files. NBSIR 86-3335. Gaithersburg, Maryland: U.S. Department of Commerce.

Semiconductor Memories

Semiconductor memories were examined as possible candidates for archival storage and were dismissed summarily because of cost, size, power, and especially volatility. Given cost projections of \$5 for a 256,000-bit CMOS random access memory (RAM) chip, the cost per bit is 2×10^{-3} cents. For large data storage systems of 10^{12} bits, the chip cost alone would be \$20 million, which is only a fraction of the total system cost.

The density of bits on a chip may be as high as 5×10^6 bits per square inch, but chip size takes only a minute portion of a semiconductor memory system. Therefore, the system would be very large. The operating power per chip is low (200 mw), but using approximately 5×10^6 chips for a large system means that 1 megawatt of power would be required. Even if cost, size, and power were not a problem, no one would want to commit archival records to a memory that would be erased if power were interrupted.

Glossary

Acid sizing.   See *Sizing.*

Acidity.   In paper, the condition that results in an acid solution when the paper is extracted with water. A detailed description is given in TAPPI test methods T428 and T509.

Alkaline fillers.   Mineral pigments of fillers such as calcium carbonate added to paper to neutralize acids that may form in the sheet. These fillers provide an alkaline reserve in the sheet.

Alkaline sizing.   See *Sizing.*

ANSI.   American National Standards Institute (a standards organization).

Archival film.   A photographic film that is suitable for the preservation of records having permanent value when stored under archival storage conditions and provided the original images are of suitable quality.

Basis weight.   The weight in grams per square meter, as described in TAPPI test method T410. Basis weight is also expressed as pounds of a ream of 500 sheets of a given size. The sheet sizes are not the same for all kinds of paper.

Compact disk.   A prerecorded 4.7–inch-diameter optical disk used to store audio data in digital form.

Computer output microfilm (COM).   Microfilm produced by a recorder that takes data directly from a computer, converts it into alphanumeric form, and records it via a camera directly onto film without any intermediate paper copy.

Cotton ginning.   The separation of the seed hulls and other small objects from the fibers of cotton.

Cross direction.   The direction in a paper or plastic web that is perpendicular to the machine or length direction.

Density.   The degree of opacity of a photographic film image, normally expressed in logarithmic terms.

Diazo film.   A type of microform in which the active ingredient is a light-sensitive diazo dye. It is used mainly for duplication and is not archivally permanent.

Elastic behavior. The behavior of a material that returns instantaneously to its initial form or state after the forces that caused its deformation are removed.

Erasable optical disk. An optical disk onto which data may be written, erased, and rewritten.

Fiber. A hollow tubular cell that gives strength and support to plant tissue. The walls of plant fibers are largely composed of cellulose.

Filler. Mineral pigments such as clay, calcium carbonate, and titanium dioxide that are added to the fiber furnish of paper.

Finding aid. A document listing, indices, or other description to aid in identifying and locating a record.

Flat-bed camera. See *Planetary camera.*

Flow camera. A camera system into which the original document is inserted and the document's continuous movement is synchronized with the film movement during exposure. Also known as a *Rotary camera.*

Formation. A property of a sheet of paper determined by the degree of distribution uniformity of the solid components with special reference to the fibers. It is usually judged by the visual appearance of the sheet when viewed by transmitted light.

Free sheet. Paper free of mechanical wood pulp or other lignin-containing pulps.

Furnish. The raw materials placed in a beater for making paper pulp.

Hygroexpansivity. The change in dimension of paper that results from a change in the ambient relative humidity. It is commonly expressed as a percentage.

Hygroscopicity. The ability to absorb water vapor from the surrounding atmosphere. It is measured by the change in moisture content with the relative humidity of the atmosphere.

Inelastic behavior. The behavior of a material that exhibits irrecoverable response to forces (compare *Elastic behavior*).

Interleaving. The placement of a document between two sheets of polyester film.

Intrinsic value. The archival term applied to permanently valuable records that are archivally acceptable only in their original physical form.

Long-term film. A photographic film that is suitable for the preservation of records for a minimum of 100 years when stored under "archival" conditions, provided that the original images are of suitable quality.

Machine direction. The direction in a paper or plastic web that is parallel to the length direction during manufacture (compare *Cross direction*).

Magneto-optical recording. An erasable technique for storing data by means of an optically addressed magnetic medium.

Medium-term film. A photographic film that is suitable for the preservation of records for a minimum of 10 years when stored under "medium-term" conditions, provided that the original images are of suitable quality (compare *Long-term film*).

Microfiche. A sheet of flat photographic film containing rows of images with an eye-legible title. Normal size is 6 × 4 in. (148 × 105 mm).

Microfilm.   Photographic film used in micrographics.

Microforms.   A term embracing both roll and flat microfilm formats.

Micrographics.   The miniaturization of images onto photographic film.

Negative film.   A photographic film in which the light and dark areas (polarity) of the film are the reverse of the original.

Optical disk.   A round flat storage medium on which information is stored that is read and/or written using focused light (usually a laser).

Orthotropic.   Having three mutually perpendicular symmetry planes: the machine, the cross-machine, and the thickness directions. Paper is an orthotropic material.

Permanence.   The property of a material that resists changes in any or all of its properties with the passage of time. Permanence is affected by temperature, humidity, light, and the presence of chemical agents such as acids. It is estimated by accelerated oven-aging tests or by tests under other specified conditions of temperature, light, and humidity.

Permanent paper.   Paper that is usually acid-free and is made to resist the effects of aging to a greater degree than is usual in other papers.

Phase change recording.   A technique for storing data by means of a laser-induced, optically differentiable localized state change. Erasability is provided if the state change is reversible.

Planetary camera.   A camera containing a flat bed on which the document to be filmed is placed and a rear-mounted central column that holds the camera head over the flat bed. Also known as a *Flat-bed camera.*

Positive film.   A photographic film in which the polarity of the film is the same as the original (compare *Negative film*).

Read only.   An information storage medium or system from which prerecorded data may only be read and not rewritten or erased.

Reduction ratio.   The linear reduction achieved by microfilming expressed as a ratio of the original. The area reduction is the square of the linear reduction.

Resolution.   A measure of the ability of photographic film to reproduce fine lines and detail. It is expressed as the number of lines to the millimeter as measured from a test target.

Rotary camera.   See *Flow camera.*

Silver halide film.   Photographic film in which the light-sensitive ingredient is a silver halide emulsion.

Sizing.   The process of adding materials to paper to render the sheet more resistant to penetration by liquids. Rosin, starches, and synthetic resins are used as sizing agents. Rosin is an example of an acid sizing, and alkaline sizing is achieved with synthetic sizes such as alkyl ketene dimers or alkenylsuccinic anhydride.

Step-and-repeat camera.   A microfiche camera of the planetary type (see *Planetary camera*) with a special step-and-repeat mechanism for filming frames along the rows in accordance with the desired format. This camera uses sheet or roll film 105 mm wide.

TAPPI.   Technical Association of the Pulp and Paper Industry (an industry standards group).

Vesicular film.   A type of photographic film in which the image consists of tiny bubbles or vesicles in a polymer binder and is developed by heat. It is used mainly for duplication and is not considered archivally permanent.

Videodisk.   A prerecorded 12-in.-diameter optical disk used to store video data in analog form.

Viscoelastic behavior.   The force-deformation or stress-strain response of a material. It depends on the material, temperature at the time of load, rate of strain, and duration of the load.

Write once.   An information storage medium or system such that once a physical data block is written it may not be erased or rewritten but only read.

Wood fiber.   Elongated, hollow cells comprising the structural units of woody plants. The term "fiber" is botanically applicable to hardwoods only, while in softwoods such cells are properly called tracheids.

Young's modulus.   The ratio of force (stress) to resulting elongation (strain) of a material; a measure of stiffness.

Biographical Sketches of Committee Members

PETER Z. ADELSTEIN received a B.Eng. in 1946 and a Ph.D. in physical chemistry in 1949 from McGill University. Upon graduation he joined the staff of Eastman Kodak, where he is now unit director, Physical Performance Section of the Materials Science and Engineering Division. He has served as chairman of Subcommittee-3 of the American National Standards Institute since 1967 and of Working Group-5 of the International Standards Organization Technical Committee-42 since 1973. He is a member of the American Chemical Society, American Society for Testing and Materials, National Fire Protection Association, and Society for Photographic Science and Engineering. His areas of interest include the behavior of high polymers in solution, physical properties of high-polymer materials in solid state, physical behavior of photographic film, and archival stability of photographic materials.

NORBERT S. BAER received his B.Sc. degree from Brooklyn College in 1959, his M.Sc. degree in physical chemistry from the University of Wisconsin in 1962, and his Ph.D. degree in physical chemistry from New York University in 1969. He was employed as a physicist at the Warner and Swasey Company from 1962 to 1963. He served as assistant professor and lecturer at Queensborough Community College from 1967 to 1969. In 1969 he joined New York University as instructor and moved up to his present position as Conservation Center Professor of the Institute of Fine Arts in 1978. In 1970 he received the Jay Krakauer Memorial Plaque and in 1983 he was a John Simon Guggenheim Memorial Foundation Fellow. His professional activities include serving as editor and board member of various publications and authoring many technical works. In 1980 he chaired the National Materials Advisory Board Committee on Conservation of Historic Stone Buildings and Monuments, and since 1980 has chaired the National Archives Advisory Committee on Preservation. He is a member of numerous professional organizations, including the Air Pollution Control Association, the American Institute for Conservation, and the International Institute for Conservation.

GLEN R. CASS received a B.S. degree in 1969 from the University of Southern California and an M.S. in 1970 from Stanford University. Both degrees were in mechanical engineering. In 1978 he received a Ph.D. in environmental engineering from California Institute of Technology. He served as an engineer at the Naval Undersea Center, Pasadena, in 1969 and as a commissioned officer at the U.S. Public Health Service from 1970 to 1973. In 1978 he became a senior research fellow and instructor at the California Institute of Technology, where he has since moved to his present position as associate professor of environmental engineering. He has served as a consultant to the Southern Coast Air Quality Management District since 1978, and between 1980 and 1984 was a member of the research screening committee of the California Air Resources Board. His areas of interest include air pollution control strategy design, aerosol mechanics, air pollution source characteristics and control technology, visibility, fluid mechanics applied to air quality problems, and environmental economics.

HANS H. G. JELLINEK received D.I.C. and Ph.D. degrees in physical chemistry from the University of London in 1941 and 1942 respectively. In 1945 he received his Ph.D. in colloid and physical chemistry and in 1964 his Sc.D. from Cambridge University. He was employed as section head in physical chemistry at J. Lyon and Company from 1945 to 1950. He was associate professor at the University of Adelaide from 1950 to 1954, visiting professor at the Polytechnic Institute of Brooklyn from 1954 to 1957, associate professor at the University of Cincinnati from 1957 to 1959, and professor of chemistry at the University of Windsor from 1959 to 1964. He was appointed professor of chemistry at Clarkson University in 1964 and since his retirement has held the title of research professor of chemistry. He served as science expert for the U.S. Department of the Army from 1954 to 1963 and also since 1970. He was a member of the committee on high-polymer research of the National Research Council of Canada from 1962 to 1964. He is a member of the American Association for the Advancement of Science and the American Chemical Society, and a fellow of the American Institute of Chemists, the Chemical Institute of Canada, and the Royal Institute of Chemistry. His areas of interest include high polymers, reaction kinetics, surface chemistry, energy production, and interfacial and rheological properties of ice.

LEON KATZ received a B.S. degree in organic chemistry from Trinity College in 1944 and a Ph.D. in organic chemistry from the University of Illinois in 1947. He was employed at GAF Corporation from 1953 to 1970 and was vice president for research and development, and later executive vice president, at Rockwood Industries from 1970 to 1972. He was vice president for commercial development at Polychrome Corporation from 1972 to 1973, and vice president for research and development in the packaging division of American Can Company from 1973 to 1982. In 1982 he joined James River Corporation, and since 1982 he has been senior vice president for corporate research and development. He served with the Army of the United States from 1943 to 1944. He is a member of the American Chemical Society, the American Association for the Advancement of Science, New York Research Directors, Technical Association of the Pulp and Paper Industry, New York Academy of Sciences, Sigma Xi, Pi Mu Epsilon, and Phi Lambda Upsilon. His areas of interest include novel packaging materials, polyelectrolytes, surfactants, specialty chemicals, fiber and paper science, and reprography.

GEORGE B. KELLY, JR., received a B.S. degree in chemistry from the University of Maryland in 1937. He was employed at Armstrong Cork Company from 1937 to 1938, Research Associates, Inc., in 1938, and Briggs Oil Clarifier from 1938 to 1942. He served in the U.S. Army and moved from captain to lieutenant colonel between 1942 and 1946. He worked at General Chemical Corporation from 1946 to 1947, Food Machinery and Chemical Corporation from 1947 to 1957, Union Carbide Corporation from 1957 to 1971, and the Library of Congress from 1971 to 1982. Since his retirement in 1982 he has served as a consultant on paper chemicals. He is a member of the Technical Association of the Pulp and Paper Industry, the American Chemical Society, and Alpha Chi Sigma. His areas of interest include floor tile research, fuel cells, fuel for space vehicles, research on chemicals to preserve paper and books, organic and inorganic chemistry, paper chemistry, physical chemistry, inorganic refining process, water-soluble polymers, and additives for paper preservation.

JOHN C. MALLINSON received B.A. and M.S. degrees in physics from Oxford University in 1953. He was employed as an engineer at AMP, Inc., from 1956 to 1961, and as senior physicist, member of the research staff in physics, and manager of the magnetics section of the research division at AMPEX Corporation from 1961 to 1984. Since 1984 he has been director of the Center for Magnetic Recording Research at the University of California, San Diego, in La Jolla. He is a member of the American Institute of Physics, the Institute of Electrical and Electronics Engineers, and the British Institute of Physics. His areas of interest include magnetic switching mechanisms in bulk and particulate materials, mechanisms of magnetic recording and magnetic field configurations, and magnetic thin film technology for recording and memory.

ERNEST R. MAY received his A.B. (1948) and Ph.D. (1951) degrees from the University of California, Los Angeles. He was a lecturer in history at Los Angeles State College in 1950 and a member of the history section of the Joint Chiefs of Staff from 1952 to 1954. In 1954 he was appointed instructor of history at Harvard University and attained the rank of professor in 1963. He was dean of Harvard College from 1969 to 1971, acting associate dean of the Faculty of Arts and Sciences from 1971 to 1972, director of the Institute of Politics from 1971 to 1974, and chairman of the Department of History from 1976 to 1979. In 1981 he was named Charles Warren Professor of History. He has authored numerous books and articles on U.S. political history, foreign policy, and historical events. He has served on several national committees and boards and is a fellow in a number of professional organizations. He is a member of the Massachusetts Historical Society, the American Historical Association, the Society of Historians on American Foreign Relations (president 1982–1983), American Association of University Professors, and the American Academy of Arts and Sciences. He chaired the Committee on the Records of Government, sponsored by the American Council of Learned Societies, the Social Science Research Council, and the Council on Library Resources, which issued its report in March 1985.

TERRY O. NORRIS received his B.S. and Ph.D. (1954) degrees in chemistry from the University of North Carolina. He worked as an analytical research chemist at E. I. duPont de Nemours and Company from 1949 to 1951 and, following an absence for graduate study, was a research chemist on polyester photographic film base and photographic coatings from 1954 to 1956. In 1956 he joined Keuffel and Esser Company as a research chemist on paper, film, and litho-plate-based products, including photographic, photohardenable, diazotype, and drafting media, and was director of research from 1958 to 1963. He worked from 1963 to 1966 for IBM Corporation as research manager of computer storage and output copying materials research, including magnetic recording, diazo microfilm, and photocopying. In 1966 he became director of research at Nekoosa Edwards Paper Company (now Nekoosa Papers, Inc., a division of Great Northern Nekoosa Corporation), and in 1969 he assumed his current position of vice president of research and development. He is immediate past president of the Technical Association of the Pulp and Paper Industry and is a member of a number of other scientific and professional organizations. His areas of interest include polymer systems, coating materials, light-sensitive systems, electrophotography, recording materials technology, pulp and paper chemistry, and paper coatings.

TED F. POWELL received a B.S. degree in accounting from the University of Utah in 1962. He worked as comptroller for Magic Chemical Company from 1962 to 1966, was a partner in the Beacon Insurance Agency from 1966 to 1967, and in 1967 joined the Genealogical Society of Utah, where he is now director of the micrographics division. He is a member of the International Council of Archives, the Committee on Reprography, and the Association for Information and Image Management, and a former member of the Society of American Archivists. His areas of interest include library science, genealogical library microfilming techniques, archiving decisions, assessment of genealogical files, and methods of records preservation.

KWAN Y. WONG received a B.S. degree in 1960 and an M.E. degree in 1963 from the University of New South Wales and a Ph.D. (electrical engineering) in 1966 from the University of California, Berkeley. In 1966 he joined the IBM Research Laboratory in San Jose, where he is now manager of the engineering and special systems department. He is a member of the Institute of Electrical and Electronics Engineers and Sigma Xi. His areas of interest include image processing, data compression, and pattern recognition.

FRANCIS T. S. YU received a B.S.E.E. degree in 1956 from Mapus Institute of Technology (Philippines) and M.S.E. (1958) and Ph.D. (1964) degrees from the University of Michigan, all in electrical engineering. He worked as a teaching fellow, instructor, and lecturer in the Electrical Engineering Department and as a research assistant in the Communications Sciences Laboratory of the University of Michigan from 1958 to 1965. In 1966 he was appointed professor of electrical and computer engineering at Wayne State University, where he remained until 1979, when he accepted the position of professor in the electrical engineering department at Pennsylvania State University. He received the 1983 Faculty Scholar Medal and was the Outstanding Researcher in the College of Engineering in 1984. In 1985 he was named Evan Pugh Professor of Electrical Engineering. He is a consultant to several industrial and government laboratories and is the author of four books as well as numerous technical articles.

Index

Photo Credits

Glen R. Cass, Environmental Quality Laboratory, California Institute
    of Technology: page 10
Genealogical Society of Utah: pages 14, 48
Library of Congress: page 85
John C. Mallinson, Center for Magnetic Recording Research,
    University of California, San Diego: page 60
National Security Agency: pages 67, 70
Nekoosa Papers Inc.: page 36
Yoichi R. Okamoto: page 40
Hugh Talman, National Archives: cover, frontispiece, pages xviii, xx,
    3, 4, 7, 17, 32, 43, 50, 54, 77, 78